Food and Agricultural Wastewater Utilization and Treatment

Food and Agricultural Wastewater Utilization and Treatment

Second Edition

Sean X. Liu
US Department of Agriculture, Agriculture Research Service,
National Center for Agricultural Utilization Research, USA

WILEY Blackwell

This edition first published 2014 © 2014 by John Wiley & Sons, Ltd

Registered office: John Wiley & Sons, Ltd, The Atrium, Southern Gate, Chichester, West Sussex, PO19 8SQ, UK

Editorial offices: 9600 Garsington Road, Oxford, OX4 2DQ, UK
The Atrium, Southern Gate, Chichester, West Sussex, PO19 8SQ, UK
111 River Street, Hoboken, NJ 07030-5774, USA

For details of our global editorial offices, for customer services and for information about how to apply for permission to reuse the copyright material in this book please see our website at www.wiley.com/wiley-blackwell.

The right of the author to be identified as the author of this work has been asserted in accordance with the UK Copyright, Designs and Patents Act 1988.

All rights reserved. No part of this publication may be reproduced, stored in a retrieval system, or transmitted, in any form or by any means, electronic, mechanical, photocopying, recording or otherwise, except as permitted by the UK Copyright, Designs and Patents Act 1988, without the prior permission of the publisher.

Designations used by companies to distinguish their products are often claimed as trademarks. All brand names and product names used in this book are trade names, service marks, trademarks or registered trademarks of their respective owners. The publisher is not associated with any product or vendor mentioned in this book.

Limit of Liability/Disclaimer of Warranty: While the publisher and author(s) have used their best efforts in preparing this book, they make no representations or warranties with respect to the accuracy or completeness of the contents of this book and specifically disclaim any implied warranties of merchantability or fitness for a particular purpose. It is sold on the understanding that the publisher is not engaged in rendering professional services and neither the publisher nor the author shall be liable for damages arising herefrom. If professional advice or other expert assistance is required, the services of a competent professional should be sought.

Library of Congress Cataloging-in-Publication Data

Liu, Sean X.
 Food and agricultural wastewater utilization and treatment / by Sean X. Liu. – Second edition.
     pages cm
 Includes bibliographical references and index.
 ISBN 978-1-118-35397-4 (cloth)
 1. Agricultural wastes – Management. 2. Sewage – Purification. I. Title.
 TD930.L578 2014
 628.1′684 – dc23
                                      2013046773

A catalogue record for this book is available from the British Library.

Wiley also publishes its books in a variety of electronic formats. Some content that appears in print may not be available in electronic books.

Cover image: Image credit: © iStock/GAPS Fotografie
Cover design by Meaden Creative

Typeset in 10/12pt Times-Roman by Laserwords Private Limited, Chennai, India.
Printed and bound in Singapore by Markono Print Media Pte Ltd

1 2014

Contents

Preface to the second edition   ix

1   Introduction   1
    1.1   Characteristics of agricultural and food wastewater   1
    1.2   Material balances and stoichiometry   18
    1.3   Fluid flow rate and mass loading   20
    1.4   Kinetics and reaction rates   20
    1.5   Theoretical modeling and design of biological reactors   24
    1.6   Process economics   27
    1.7   Further reading   28
    1.8   References   28

2   Basic microbiology in wastewater treatment   29
    2.1   Introduction   29
    2.2   Structures of cells   30
    2.3   Important microorganisms in wastewater   30
    2.4   Microbial metabolism   37
    2.5   Nitrification   39
    2.6   Denitrification   41
    2.7   Further reading   42
    2.8   References   43

3   Physicochemical wastewater treatment processes   45
    3.1   Introduction   45
    3.2   Equalization basins   46
    3.3   Screening   48
    3.4   Flotation   49
    3.5   Sedimentation   52
    3.6   Coagulation and flocculation   57
    3.7   Filtration processes   62
    3.8   Adsorption   65
    3.9   Chemical oxidation   70
    3.10   Membrane separations   71
    3.11   Ion exchange   93
    3.12   Closing remarks   99

	3.13	Further reading	101
	3.14	References	101

4 Biological wastewater treatment processes — 103

4.1	Introduction	103
4.2	Kinetics of biochemical systems in wastewater microbiology	104
4.3	Idealized biochemical reactors	110
4.4	Completely mixed aerated lagoon (CMAL)	114
4.5	Trickling filter (TF)	116
4.6	Rotating biological contactor (RBC)	119
4.7	Combined aerobic processes	125
4.8	Contact anaerobic systems	126
4.9	Further reading	131
4.10	References	131

5 Advanced wastewater treatment processes — 133

5.1	Introduction	133
5.2	Biological removal of nitrogen: nitrification and denitrification	135
5.3	Physicochemical removal of nitrogen	136
5.4	Biological removal of phosphorus	137
5.5	Physicochemical removal of phosphate	141
5.6	Membrane processes for advanced wastewater treatment	143
5.7	VOC removal with pervaporation	146
5.8	Disinfection	147
5.9	Further reading	150
5.10	References	150

6 Natural systems for wastewater treatment — 153

6.1	Introduction	153
6.2	Stabilization ponds	154
6.3	Land treatment systems	166
6.4	Wetland systems	181
6.5	Floating aquatic plant systems	185
6.6	Further reading	192
6.7	References	192

7 Sludge treatment and management — 195

7.1	Sludge quality and characteristics	196
7.2	Sludge thickening (concentration)	197
7.3	Sludge stabilization	201
7.4	Reed beds	205
7.5	Conditioning of sludge	208
7.6	Dewatering	209
7.7	Land applications and surface disposal	218
7.8	Incineration	219
7.9	Further reading	222
7.10	References	222

8 Recoverable products and energy from food and agricultural wastewaters   225

8.1 Introduction   225
8.2 Water recovery and reuse in the food and agricultural processing industries   228
8.3 Recoverable carbohydrates, fats, and proteins for human and animal consumption   229
8.4 Recoverable aroma flavoring compounds from food processing   232
8.5 Recoverable food/agricultural biomaterials for non-food uses   233
8.6 Energy or fuel generations from wastewaters   234
8.7 Algae-based biodiesel and fuel ethanol   237
8.8 Potential applications of industrial commodities derived from sludge treatment   239
8.9 Further reading   242
8.10 References   242

9 Economics of food and agricultural wastewater treatment and utilization   247

9.1 Introduction   247
9.2 Estimating the unit cost of treating food and agricultural wastewater   248
9.3 Estimating overall costs of wastewater treatment processes with substance and energy recovery   253
9.4 Further reading   254
9.5 References   254

Index   255

Preface to the second edition

The first edition of *Food and Agricultural Wastewater Utilization and Treatment* was published in 2007 by Blackwell Scientific Publishers in Ames, Iowa, USA. Since then, Blackwell Scientific Publishers has been acquired by Wiley (now part of Wiley-Blackwell). I firmly believe that this book has been of interest to those in the food and agricultural processing businesses and will help them to deal with nutrient-rich effluents from every stage of their businesses. Many have bought the first edition of this book, and several have provided positive reviews. I am heartened to know that many people have found this book useful, and feel that a revision and expansion of the book will serve the community better. The new edition's target audience is again anyone who has, or potentially has, the responsibility of dealing with plant effluents in agricultural and food processing operations. It should be of value to entrepreneurs who view wastewaters from food and agricultural processing as untapped sources of wealth and opportunities to "do things big".

The world economy has profoundly changed since the first edition of this book in 2007, and so has the field of environmental protection. I have revised the book throughout the new edition, reflecting this new reality; in addition to correcting typographic errors, I added new information and updated the old. Chapter 8 has been expanded to include many new developments in the field of waste utilization. As the prices of agricultural commodities continue to hover around an all-time high, the future of agricultural and food wastewater utilization has never been better.

The book, like its predecessor, has not been written in the form of a plain vanilla college textbook, although it certainly can be a good reference book for college seniors or graduate students working either on theses or on projects in the field of environmental science and engineering. Rather, it is, designed to fill a void that has long been apparent and will remain so for years – a *vade mecum* for everyone who is interested in the fields of agricultural and food wastewater treatment and utilization.

1
Introduction

1.1 Characteristics of agricultural and food wastewater

Whenever and wherever food in any form is handled, processed, packed and stored, there will always be unavoidable generation of wastewater. Wastewater is the most serious environmental problem in the manufacturing and processing of foods. Most of the volume of wastewater comes from cleaning operations at almost every stage of food processing and transportation operations. The quantity and general quality (i.e., pollutant strength, nature of constituents) of this generated processing wastewater have both economic and environmental consequences with respect to its treatability and disposal.

The cost for treating the wastewater depends on its specific characteristics. Two significant characteristics that dictate the cost for treatment are the daily volume of discharge and the relative strength of the wastewater. Other characteristics become important as system operations are affected and specific discharge limits are identified (e.g., suspended solids). The environmental consequences in inadequate removal of the pollutants from the waste stream can have serious ecological ramifications. For example, if inadequately treated wastewater were to be discharged to a stream or river, a eutrophic condition might develop within the aquatic environment due to the discharge of biodegradable oxygen-consuming materials. If this condition were sustained for an extended period of time, the ecological balance of the receiving stream, river or lake (i.e., aquatic microflora, plants and animals) would be upset. Continual depletion of the oxygen in these waters would also give rise to the development of obnoxious odors and unsightly scenes.

Knowledge of the characteristics of food and agricultural wastewater is essential to the development of economical and technically viable wastewater management systems that are in compliance with current environmental policy and regulations. Management methods that may have been adequate with other industrial wastewaters may be less feasible with food and agricultural wastewater, unless the methods are modified to reflect the characteristics of the

---

Food and Agricultural Wastewater Utilization and Treatment, Second Edition. Sean X. Liu.
© 2014 John Wiley & Sons, Ltd. Published 2014 by John Wiley & Sons, Ltd.

Table 1.1 Wastewater treatment options available to remove various categories of pollutants in food and agricultural wastewater

Pollutants in wastewaters	Management options
Dissolved organic species	Biological treatment; adsorption; land applications; recovery and utilization.
Dissolved inorganic species	Ion exchange; reverse osmosis; evaporation/distillation; adsorption.
Suspended organic materials	Physicochemical treatment; biological treatment; land applications; recovery and utilization.
Suspended inorganic materials	Pretreatment (screen); Physicochemical treatment (sedimentation, flotation, filtration, coagulation).

wastewater and the opportunities it may hold. The wastewaters produced in agricultural processing and food processing vary in quantity and quality, with those streams from food processing typically having low strength and high volume, while those coming from animal farming operations tend to be high in strength but low in volume. These differences in quantity and quality dictate the type and capacity of wastewater management systems that should be deployed.

A clear understanding of the characteristics of food and agricultural wastewater permits management decision on treatment and utilization methods that are effective and economical, and this point is further spelt out in Table 1.1. For example, a low-strength and high-volume wastewater containing small amount of organic colloidal particulates may require a stand-alone biological wastewater treatment facility or a just a frame-and-plate filter press; the decision is both technical and economical. Another generalized observation is that the bulk oxygen-demanding substances are in the liquid phase for food processing wastewater, while most oxygen-demanding substances in the wastewater of a high-intensity livestock farming operation are in the form of solid particulates.

Some food processing operations occur seasonally (e.g., processing of fruits and vegetables). This seasonality adds complexity to the wastewater management systems that handle different sources of food and agricultural wastewater year round and, clearly, the understanding of wastewater characteristics helps plan ahead for such process operations. Knowledge of wastewater characteristics also allows strategic planning of water recycling and the reuse and recovery of valuable components in the wastewater.

As in most wastewaters, the components present in agricultural and food wastewater run a gamut of many undefined substances, almost all organic in nature. Organic matters are substances containing compounds comprised mainly of the elements carbon (C), hydrogen (H) and oxygen (O). The carbon atoms in the organic matter (also called carbonaceous compounds) may be oxidized both chemically and biologically to yield carbon dioxide (CO_2) and energy.

It is possible that some sources of wastewaters from certain food processing operations in a processing plant may have limited numbers of possible contaminants present. However, these wastewaters tend to mix with other streams of wastewaters from the same work site, making it virtually impossible to catalog all the substances in the effluents from the plant. Thus, the characteristics of agricultural and food wastewater can be viewed as a set of well-defined physicochemical and biological parameters that are critical in designing and managing agricultural and food wastewater treatment facilities.

1.1.1 General characteristics of wastewaters in agriculture and food processing

Wastewater from food processing operations is defined by the food itself. Food and agricultural wastewater contain dissolved organic solids from various operations and debris from mechanical processing of foods, such as peeling and trimming and hydrodynamic impacts in washing and transporting. Agricultural and food processing operations inevitably use large quantities of water to wash and, in some instances, cool food items. Canning wastewaters are essentially the same as home kitchen waste, as the wastewater is accumulated from various processes involved in the canning operations, such as trimming, sizing, juicing, pureeing, blanching, and cooking. Blanching of vegetables also requires large amounts of water to blanch and cool blanched vegetables. Almost all operation in food or agricultural processing involves cleaning of plant floors, machinery, and processing areas, often mixed with detergents that sometimes double as lubricants for the food processing machinery.

Depending on particular processing operations, water used in the operations is often reused, with or without treatment, when such practice is economical and legal. As fresh water supply is limited in many parts of the world, reusing water is often seen as a must for practical reasons. The reuse and recycling of water can result in considerable reduction in water usage. However, one should keep in mind that if the reused water is intended for edible food items, food safety issues arising from the reused water should be examined diligently and thoroughly. After all, food safety issues remain the overriding concern in all food processing and manufacturing operations.

Common pollutants present in the majority of food and agricultural wastewater and effluents from each stage of the typical wastewater treatment processes (see the following chapters for more information) include free and emulsified oil/grease, suspended solids, organic colloids, dissolved inorganic, acidity or alkalinity, and sludges. Table 1.1 is a summary of the processes available to treat food and agricultural wastewater.

Each food processing plant produces wastewaters of different quantity and quality. No two plants, even with similar processing capacity of food products, will generate wastewaters of the same quantity and quality, since there are too many variables (technical or otherwise) in the processes that ultimately define characteristics of wastewater. Furthermore, even different periods of food processing in the same plant may produce different wastewater streams with different characteristics. It is, therefore, essential to understand that the generalized description of wastewaters from fruit and vegetable processing needs to be understood as an approximate explanation of a complex issue. Any quantitative information shown here or anywhere else must be considered as averaged data. Typical characteristics, estimated volume, and estimated organic loading of wastewater generated by the food processing industry in the state of Georgia, USA, are shown in Table 1.2.

All major food and agricultural processing operations generate wastewater streams. However, the amount and strength of the wastewater streams varies with the major segments of the food and agricultural processing industry. Table 1.3 summarizes the sources of the wastewater streams and possible treatment processes.

As shown in Table 1.4, not all agro-food processing operations generate wastewater in such substantial quantities as to warrant on-site wastewater treatment facilities.

The following summary of the major segments of the agro-food processing operations requiring wastewater treatment is presented for the reader to appreciate the unique pollution issues therein, even though it is clear that there is considerable similarity among many segments of the food and agricultural processing industry. Additional information about the characteristics of wastewaters in all major segments of the food and agricultural processing industry can be found from Middlebrooks (1979).

Wastewaters from fruit and vegetable processing

The fruit and vegetable industries are as assorted as the names imply; these industries process the great variety of fruits and vegetables grown in the United States in a number of ways. The categories of processing include canning, freezing, dehydrating, and pickling and brining. The quantity and quality of wastewater streams from the industries vary considerably with the operations of the processing and the changing seasons.

Fruit and vegetable processing plants are major water users and waste generators. In all stages of food processing (unitary processes), raw foods must be

1.1 CHARACTERISTICS OF AGRICULTURAL AND FOOD WASTEWATER

Table 1.2 Typical characteristics, estimated volume and estimated organic loading of wastewater generated by the food processing industry in Georgia, USA

Industry group	Estimated wastewater volume, million gallons/year	Typical characteristics	Estimated organic loading, tons/year BOD
Meat and poultry products	10,730	1,800 mg/L BOD 1,600 mg/L TSS 1,600 mg/L FOG 60 mg/L TKN	80,600
Dairy products	500	2,300 g/L BOD 1,500 mg/L TSS 700 mg/L FOG	14,900
Canned, frozen and preserved fruits and vegetables	2,080	500 mg/L BOD 1,100 mg/L TSS	4,300
Grain and grain mill products	130	700 mg/LBOD 1,000 mg/L TSS	300
Bakery products	530	2,000 mg/L BOD 4,000 mg/L TSS	4,400
Sugar and confectionery products	140	500 mg/L BOD	300
Fats and oils	350	4,100 g/L BOD 500 mg/L FOG	7,000
Beverages	3,660	8,500 mg/L BOD	91,000
Miscellaneous food preparations and kindred products	700	6,000 mg/L BOD 3,000 mg/L TSS	5,600
TOTAL	18,810		208,600

Abbreviations: TSS – total suspended solids; FOG – fats, oils, and grease.
Source: Magbunua (2000). Reproduced with permission of University of Georgia, College of Engineering, Outreach Service.

rendered clean and wholesome, and food processing plants must be maintained in a sanitary condition all of the time. Several common unit operations of fruit and vegetable processing that generate wastewater are shown in Figure 1.1.

Some of these unit operations shown in Figure 1.1 are intuitively obvious generators of waste (e.g., washing and rinsing), while others are less so (e.g., in-plant transport). Table 1.3 provides a brief explanation of several unitary processes that generate wastewater. For the most part, these wastewaters have been shown to be biodegradable, although salt is not generally removed during the treatment of olive storage or processing brines, cherry brines, and sauerkraut brines.

The effluents from fruit and vegetable processing operations consist mainly of carbohydrates such as sugars, starches, pectins, and other components of the cell walls that have been severed during processing. Of the total organic

Table 1.3 Summary of wastewater sources in major food and agricultural processing

Agro-food operations	Sources of wastewater streams	Treatment strategies
Vegetables and fruits	Sorting, trimming, washing, peeling, pureeing, in-plant transport, canning and retort, dehydration, and cleanup	Primary and secondary treatment processes
Fishery	Eviscerating, trimming, washing, pre-cooking, canning and retort, and cleanup	Primary and secondary treatment processes
Poultry and meat	Animal waste, killing and bleeding, scalding (poultry), eviscerating, washing, chilling, and cleanup	Primary and secondary treatment processes
Dairy	By-products, spills, leaks, line cleaning, and cleanup	Biological wastewater treatment
Corn wet milling	Steeping water, washing, and cleanup	Mainly screen, activated sludge processes, and secondary sedimentation
Sugar refining	Process water and cooling water	Recycling and discharge to municipal wastewater systems
Oil and fat	Steaming, solvent recovery, degumming, soapstock water, neutralization, and cleanup	Primary, secondary treatment, and sludge treatment processes
Non-alcoholic beverage	Cleanup	Discharge to municipal wastewater systems
Alcoholic beverage	Washing, cooling, leaks, and cleanup	Biological wastewater treatment and stabilization ponds
Flavoring extracts	Washing, evaporator condensate, steam distillation, and cleanup	Biological wastewater treatment or direct discharge to municipal wastewater systems
Egg product	Washing, leaks, and cleanup	Biological wastewater treatment and aerobic lagoon
Other food production	Leaks and cleanup	Depending on specific products and locality

matter, 70–80% is in the dissolved form and is not easily removed from wastewater by conventional mechanical means, although physicochemical processes may be used, such as adsorption and chemical oxidation or membrane-based technologies such as membrane filtration (see Chapter 3 for adsorption, chemical oxidation and membrane filtration). Obviously, biological wastewater treatment methods will work best in this type of wastewater streams.

Table 1.4 Common unitary processes of fruit and vegetable processing that generate wastewater

Process	Wastewater comes from ...
Washing and rinsing	The entire process; may use detergent or chlorinated water.
Sorting (grading)	Density grading operation only.
In-plant transport	Water conveys products from one location to the other.
Peeling	Hot water or high-pressure water spray; may involve chemicals (caustic soda) or detergents.
Pureeing and juicing	Condensated evaporated water.
Blanching	Hot water or steam for blanching.
Canning and retort	Washing cans and steam for retort and cooling with water.
Drying or dehydration	Condensated evaporated water.
Mixing and cooking	Leaking of liquid products.
Clean-up	Cleaning up at every stage.

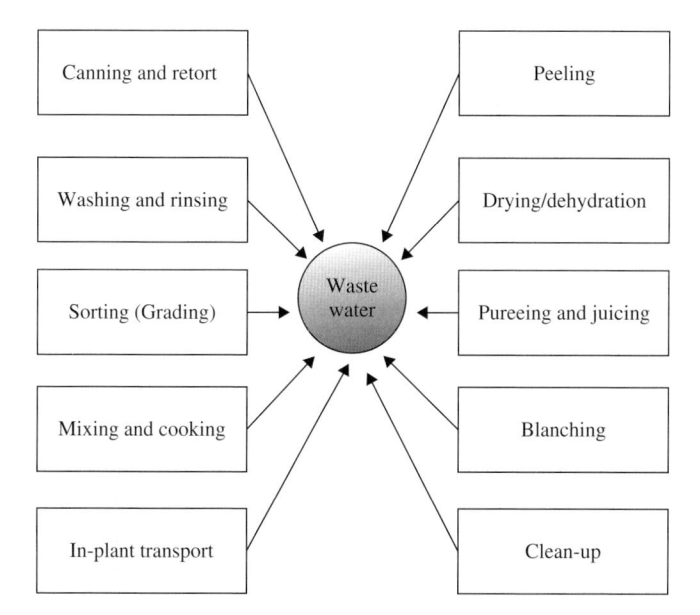

Figure 1.1 Unitary processes of fruit and vegetable processing that generate wastewater

The majority of the literature review regarding characterization of fruit and vegetable processing wastewaters focuses on wastewater streams from canning of fruits and vegetables (e.g., Soderquist *et al.*, 1975); the wastewaters from other processing operations of fruits and vegetables are of importance as well. Blanching of vegetables for freezing is a process that requires a large amount of water,

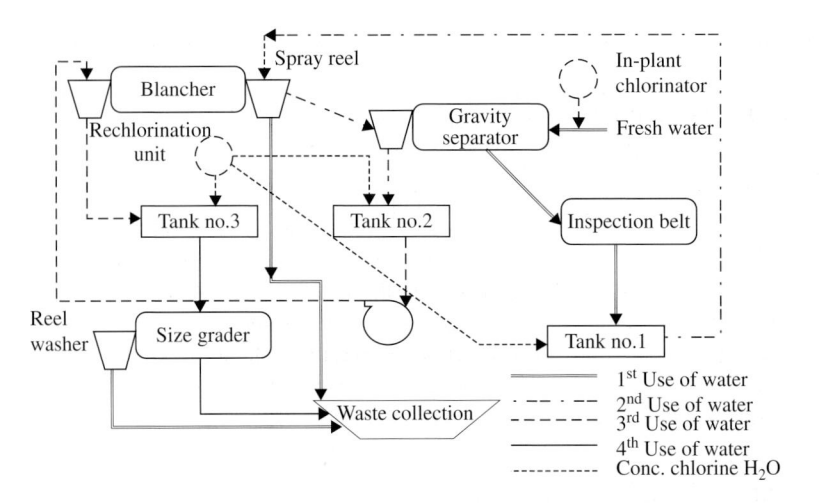

Figure 1.2 Diagram of a four-stage counterflow system for re-use of water in a pea cannery

and the quantity of wastewater generated is also proportionally high. Figure 1.2 shows a flow diagram of water reuse in a pea processing company.

Post-harvesting agricultural wastewater could also be a source of wastewater. Washing and rinsing waters used in cleaning fresh produces and fruits are sometimes reused, but wastewater is still generated in the process and has to be treated eventually. There is a possibility of recovering valuable substances from wastewater streams in fruit and vegetable processing, such as flavors from blanching waters. However, doing so is often technically complex and it may be economically impractical to extract these valuables from among a large number of undesirables in these streams using the technologies currently available.

Wastewaters from the fishery industry

The production processes used in the fishery industry generally include the following: harvesting; storing; receiving; eviscerating or butchering; pre-cooking; picking or cleaning; preserving; and packaging. Harvesting provides the basic raw materials (fish) for processing and subsequent distribution to the consumer. Once the fish are aboard the fishing vessel, the catch either is taken directly to the processor, or is iced or frozen for later delivery. Pre-processing may be carried out on board before the catch is sent to the processing plant. This may include beheading shrimp at sea, eviscerating fish or shellfish at sea, and other operations to prepare the fish for butchering. Wastes from the butchering and evisceration that are sizable are usually collected in dry form, or screened from the wastewater stream and processed as a fishery by-product.

1.1 CHARACTERISTICS OF AGRICULTURAL AND FOOD WASTEWATER

The receiving operation usually involves unloading the vessel, weighing, and transporting by conveyor or suitable container to the processing area. The catch may be processed immediately or transferred to cold storage.

Sometimes, cooking or pre-cooking of crab and other shellfishes or tuna may be practiced in order to prepare the fish or shellfish for removal of meat and cleaning operation. The inedible fish or seafood parts, such as skin, bone, gills, shell, and similar, are easily removed after pre-cooking. The steam condensate, or stick water, from tuna or crab precook is often collected and further processed as a by-product. Wastes generated during this procedure are sometimes collected and saved for by-product processing. Depending on the species of seafoods, the cleaning operation may be either manual or mechanical.

With fresh fish or fresh shellfish, the meat product is packed into a plastic container and refrigerated for shipment to a distribution center or directly to a retail outlet. If shelf life of the product is required for an extended period of time before consumption, preservation techniques must be used to prevent spoilage from bacterial activities and enzymatic autolysis. Freezing, canning, pasteurization, drying, and refrigeration are the commonest preservation techniques used in the fishery industry.

Characteristics of fishery wastewaters are often dependent of several factors, including method of processing (mechanical or manual), fish species, and fish products. However, even with similar processing plants, using the same method of processing on the same species of fish and producing the same fish products, the quality of wastewaters (in terms of BOD, COD, TOC or TSS which will be explained later) varies with location and even with season. It should be mentioned that there is no substitute for direct determination of the quality of fishery wastewater in the effluent being investigated.

Like all wastewaters under consideration for treatment, the issue of treatability of seafood or fishery wastewater is often shaped by discharge limits set up by government agencies or an international body enforced through international treaties. Specifically, the discharge limits of BOD_5, Total Suspended Solids (TSS), and fat/oil/grease (FOG) are enforced based on the variety of fish species. Table 1.5 is a summary of discharge limits imposed by US EPA in 1985. It is prudent to consult with the local authorities on issues related to discharge limits of fishery wastewaters.

Fishery wastewaters are rich in fats and proteins. According to Middlebrooks (1979), a processing plant for finfish processing can produce 3.32 kg/ton of BOD, 0.348 kg/ton of grease/oil, and 1.42 kg/ton of suspended solids in the wastewater if using manual processing, or 11.9 kg/ton of BOD, 2.48 kg/ton grease/oil and 8.92 kg/ton of suspended solids in the wastewater if using mechanical means. This has generated a lot of interest in recovering these materials to offset totally or partially the costs of treating the fishery wastewater.

Table 1.5 Summary of discharge limits for the fishery industry imposed by the United States Environmental Protection Agency in 1985

Fish species	BOD$_5$	TSS	FOG
Tuna	20.0	8.3	2.1
Salmon	2.7	2.6	0.31
Other finfish	1.2	3.1–3.6	1.0–43
Crab	0.3–10	2.2–19	0.6–1.8
Shrimp	63–155	110–320	36–126
Clam and oyster	N/A	24–59	0.6–2.4

Source: USEPA.

Like proteins, the presence of fat/oil/grease (FOG) in the fishery wastewaters is mainly due to the processing of fishes. Canning, for example, generates grease and oil after fish products are heated.

Wastewaters from meat and poultry processing

The meat and poultry processing industry (excluding rendering but including seafood processing) uses an estimated 150 billion gallons of water annually. Although a portion of the water used by the industry is reused or recycled, most of it becomes wastewater. Similar to those wastewater streams from the fishery industry, the wastewaters from meat and poultry processing are high in fat/oil/grease and proteins.

The poultry industry handles billions of kilograms of chickens (called broilers, and with weights ranging from 1.1 kg to 2.0 kg) and turkeys each year, and processing plants vary, ranging from 50,000 birds to 250,000 birds per day. The main poultry operations involve receiving and storing, slaughtering, de-feathering, evisceration, packing, and freezing. Nearly all these operations involve using water, and a great deal of pollutants in the wastewater stream are created in the receiving and storing operation, where manure and unconsumed feed are washed down from the broilers. The water usage and wastewater generation is illustrated in Figure 1.3.

A meat processing plant consists of a slaughterhouse and/or a packing house. The slaughtering process has four basic operations: killing; hide removal/hog de-hairing; eviscerating/trimming; and cooling of carcasses (US EPA, 1974). Each of these operations contributes to the wastewater stream but, before being herded to their final destinations, the animals are held in the livestock holding pens, which generates additional wastewater streams. The wastewater streams from these holding pens primarily come from spillage from the water troughs, from cleanup, and from wastes laid by the animals.

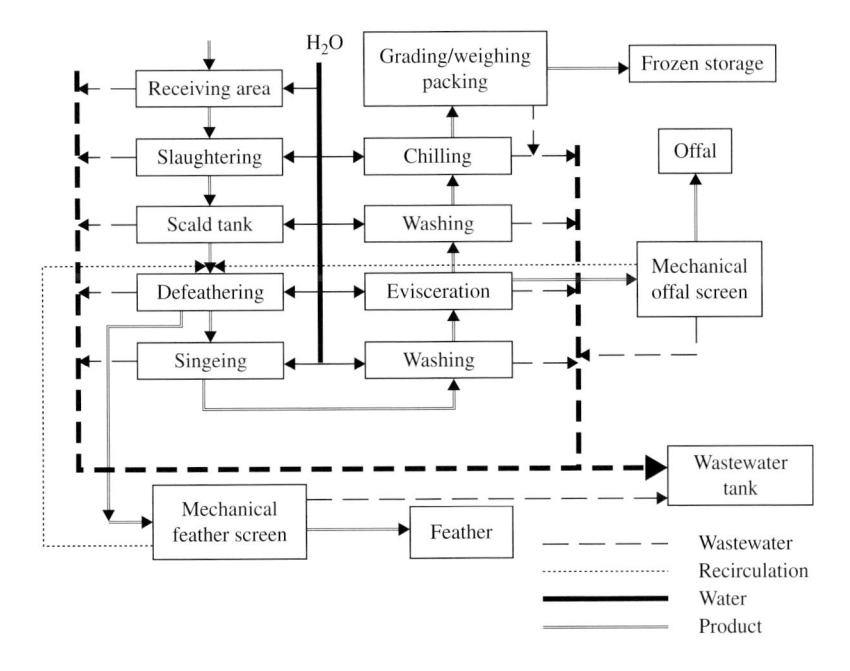

Figure 1.3 Flow chart of a poultry processing plant

Wastewaters from the dairy industry

The dairy industry is one of the most important agricultural processing industries in the United States, and it has grown steadily in recent decades.

Wastewaters originate from two major dairy processes – not only from fluid milk at the receiving station and bottling plants but, increasingly and more importantly, at the processing plants that produce condensed milk, powdered milk, condensed whey and other products such as dry whey, butter, cheese, cultured product, ice cream, and cottage cheese. The milk itself has a BOD_5 of 100,000 mg/L, and washing plants that produce butter and cheese may produce a wastewater with BOD_5 of 1,500–3,000 mg/L.

The dairy processing uses raw materials beyond milk and milk related materials; non-dairy ingredients, such as flavors, sugar, fruits, nuts, and condiments are utilized in manufacturing ice creams, yogurts, and flavored milks and frozen desserts. The pollutants can enter the wastewater streams through spills, leaks, and wasting of by-products. Apart from whey, which is acidic, most dairy wastewater streams are neutral or slightly alkaline, but they tend to become acidic rapidly due to the lactic acid produced as a result of the fermentation of lactose.

Most dairy product processing operations are multi-product facilities. Among these operations there may be receiving stations, bottling plants, creameries, ice cream plant, and cheese making plants; all these may contribute to wastewater streams in the dairy industry. Controlled products loss and recovery of by-products (e.g., whey protein) can improve not only yields (and thus profits), but also the amount and strength of dairy wastewater streams.

Dairy wastewaters are amendable to biological wastewater treatment, and this is the principal method used in the dairy industry. According to US EPA (1974), there were 64 activated sludge plants, 34 trickling filters, six aerobic lagoons, one stabilization pond, four combined systems, two anaerobic digestion facilities and one sand filtration for secondary effluent operating in the dairy industry in the United States. Most dairy processing plants treat their wastewaters to a level that is acceptable to municipal wastewater treatment facilities.

Wastewaters from oil and fat processing

Edible oil extraction involves solvent extraction of oil-bearing seeds or animal fats (there are mechanical expressers for olive oil and sesame oil) and refining steps of removing undesirables from extracted oil. In addition to cleanup and washing operations that use water, thus generating wastewater, several other processes all contribute to wastewater streams in edible oil production plants. These include deodorization that involves the injection of steam, refining that involves removing free fatty acids, phosphatides and other impurities with caustic soda, and oil recovery from the extracted meal using water.

The wastewaters from the oil production and refining industry, without doubt, are amendable to biological wastewater treatment. There are several pollutants in the wastewater streams from the edible oil extraction and refining, namely free and emulsified oils, grease, suspended solids, dissolved organic and inorganic solids. Along with the sludges that come from either primary or secondary treatment processes, many common wastewater treatment processes may be employed to remove these pollutants. Trace amounts of solvent such as hexane may be removed by adsorption or steam/air stripping. Another environmentally friendly method of removing hexane from wastewater is pervaporation (Peng *et al.*, 2003).

1.1.2 Parameters for physicochemical treatment of wastewater

pH

pH is a measurement of the acidity of the wastewater and an indication of growth conditions for the microbial communities used in biological wastewater treatment regimens. pH values vary greatly with the sources of agricultural and food wastewater, and also with the environmental conditions and duration of storage of the wastewater collected, as these factors dictate the amounts of certain substances and decomposition of biological matters, as well as emissions of ammonia compounds.

Solids content

Solids in wastewaters come in two forms: suspended solids (non-dissolvable) and dissolved solids. Suspended solids are nuisances, because they can either settle on the bottom of the receiving water body or float on the surface of the water body. Either way will affect the ecological balances of the receiving water body. Solids that readily settle are usually measured with an Imhoff cone (see Figure 1.1). Here, a known amount of water sample is poured into the cone and the amounts of the solids settled at given times are recorded and compared with the admissible amounts of settling solids in the wastewater for discharge. The acceptable settling solids level is usually determined by environmental regulations and, as a rule of thumb, discharge of wastewater or treated wastewater is not acceptable if the result of Imhoff testing shows that the water sample contains settling solids after ten minutes of testing.

Suspended solids are usually measured with a porous fiberglass filter of known pore size, in which a known amount of well-mixed water sample passes through. The dry mass accumulation on the filter is the amount of non-dissolvable solids.

Oils and greases represent another realm of suspended solids. These floating substances from some food operations have tendency to clog pipes and stick to the surfaces of any material. They are also easily oxidized, producing objectionable odors. In any case, oils and greases should be removed. The amount of oil and grease may be measured with the solvent extraction method found in the standard methods (Eaton *et al.*, 2005).

Soluble solids are laboratory measured, with evaporation and subsequent weighing of remaining dry mass of a known amount of water filtrate sample that is collected from the suspended solids measurement, or similar pre-treatment to remove suspended solids. Soluble solids are significant in some sources of food wastewater (e.g., fishery, dairy industries) and, thus, they are important in formulating wastewater treatment and resource recovery strategy.

Temperature

It is generally accepted that the temperature of wastewater discharged to a receiving water body cannot exceed 2–3°C of the ambient temperature, in order to maintain population balance of aquatic ecosystem of the receiving water body.

Wastewater from some food operations, such as retort, should be cooled before discharge or biological treatment.

Odor

Odor by itself is not a pollutant, although prolonged and intense exposure has been attributed to adverse effects on wastewater treatment plant workers and even residents living near the plant, with symptoms such as headache and nausea. Food wastewater contains significant amounts of organic matter and, when this organic matter decomposes into volatile amines, diamines and, sometimes, ammonia or hydrogen sulfide, odor results and it can be overwhelming.

The other source of wastewater odor generated in food processing is the blanching operation of certain sulfur-rich vegetables, such as cauliflowers and cabbage. The incentive for developing odor abatement strategy in food and agricultural wastewater management is obvious. The public perception and acceptance of a food processing plant are influenced often by nostril, not nostalgia.

1.1.3 Parameters for biological treatment of wastewater

The organic matters in food and agricultural wastewater are considerable and complex. Instead of attempting to identify each organic component of wastewater, wastewater professionals use the parameters for biological wastewater treatment to classify the organic materials. The most common parameters are the oxygen demand values. The term "oxygen demand" refers to the amount of oxygen that is needed to stabilize the organic content of the wastewater. The two most common oxygen demand methods of defining organic matters in wastewater are the *biochemical oxygen demand* and the *chemical oxygen demand*.

Biochemical oxygen demand

Biochemical oxygen demand is also known as its acronym, BOD, and it estimates the degree of organic content by measuring the oxygen required for the oxidation of organic matter by the aerobic metabolism of microbial communities. A characteristic simple carbonaceous compound is fructose, which is oxidized as follows:

$$C_6H_{12}O_6 + 6O_2 \rightarrow 6CO_2 + 6O_2$$

The common procedures of BOD measurements are the dilution method and the respirometric method.

1.1 CHARACTERISTICS OF AGRICULTURAL AND FOOD WASTEWATER

The dilution method is the most common method in use for wastewater industry. It consists of diluting a wastewater sample with a nutrient solution (to provide essential minerals for microbial activities) according to wastewater strength, within airtight bottles that are also saturated with air (for facilitating aerobic metabolism), and measuring the dissolved oxygen at the start and periodic intervals of the analysis. A five-day period is generally used, and the BOD measured thereafter is called BOD_5. The authoritative procedures of BOD analysis can be found in the standard methods (Eaton et al., 2005).

One cautionary note for BOD analysis is that the BOD analysis involves the degradation of organic matter by a microbial population in the testing bottles. The microbial count is important for the analysis, and insufficient microbial count will underestimate the BOD. This issue is particularly critical for food wastewater analysis, because some food processing operations involving thermal processing or other sterilizations, and the wastewater generated in those operations may not have sufficient microbial count for accurate analysis of BOD in the wastewater. A possible remedy for this is constantly to measure the wastewater from those operations for a long period of time or to add the adapted "seed" of bacteria to the wastewater. The dilution method is a time-honored but time-consuming method. An upgraded version of the method involves the use of a dissolved oxygen electrode in the form of BOD_5, enabling continuous readings of the dissolved oxygen during the five-day period. Commercial products of the BOD_5 analysis instruments based on the dilution method are available.

Another phenomenon that could alter the BOD analysis result, though not occurring in all food and agricultural wastewaters, is nitrification of the wastewater. Nitrification is a biochemical process of converting organic nitrogen (e.g., proteinaceous compounds) into nitrate (Liu et al., 2003). This is an aerobic process and, therefore, uses additional oxygen. One method of inhibiting nitrification is to use inhibitive chemicals such as allyl thiourea, methylene blue or 2-chloro-6-(trichloromethyl) pyridine (Metcalf & Eddy, Inc., 2002).

The respirometric method is an alternative to the dilution method in BOD analysis. It accelerates BOD analysis by combining biochemical processes with a faster chemical reaction. The basic design of the respirometric method is the use of a continuously stirred bottle with partially filled wastewater (and a headspace), which is connected to a reservoir of alkali (usually potassium chloride) that absorbs the CO_2 generated from the degradation of organic matters in the wastewater sample (see Figure 1.2). The pressure changes in the headspace of the BOD bottle are monitored constantly for consumption in O_2 in the wastewater sample. Even with the hybrid BOD analysis methods, the BOD analysis is slow and unsuitable for process control purposes in a wastewater treatment plant. Another approach to measuring the organic content of wastewater is chemical oxygen demand, which has been developed to complement the BOD analysis.

Chemical oxygen demand

Chemical oxygen demand (COD) is a method of estimating the total organic matter content of wastewaters, and is an approach that is based on the chemical oxidation of the organic materials in the wastewater. It involves either oxidation of the organic matters by permanganate or oxidation by potassium dichromate ($K_2Cr_2O_7$). COD analysis using dichromate is the most common method today, and it is used for continuous monitoring of biological wastewater treatment systems. The value of COD for a given wastewater stream is usually higher than that of BOD_5, due to the fact that inorganic matter can also be oxidized by potassium dichromate.

It is common to correlate the values of COD to the values of BOD_5, and to use the rapid COD analysis method (about two hours) to determine the organic content of the wastewater sample. The COD test utilizes $K_2Cr_2O_7$ in boiling concentrated sulfuric acid (150°C) in the presence of a silver catalyst (Ag_2SO_4) to facilitate the oxidation. The detailed procedures of COD test can be found in the standard methods (Eaton et al., 2005). The following reaction describes the oxidation of organic carbonaceous compounds in the presence of $K_2Cr_2O_7$ and the catalyst:

$$Cr_2O_7^{2-} + 14H^+ + 6e^- \rightarrow 2Cr^{3+} + 7 H_2O$$

The COD is calculated by titrating the remaining dichromate of known amount or by spectrophotometrically measuring the Cr^{3+} ion at 606 nm (or remaining $Cr_2O_7^{2-}$ at 440 nm). Although it is more time-consuming, the titration method is more accurate than the spectrophotometry method.

A common interference in the COD testing is chloride in the wastewater, which is readily oxidized by dichromate:

$$Cr_2O_7^{2-} + 14H^+ + 6Cl^- \rightarrow 3 Cl_2 + 2Cr^{3+} + 7 H_2O$$

This interference that causes the COD level in the wastewater to be overestimated may be prevented with the addition of mercuric sulphate ($HgSO_4$) to remove Cl^- in the form of an $HgCl_2$ precipitate (Bauman, 1974). The above COD method is called "open reflux method" in the standard methods (Eaton et al., 2005). Another COD testing method is called the closed reflux method (Eaton et al., 2005). In this setting, the oxidation takes place in closed tubes filled with a small wastewater sample mixed with Ag_2SO_4 and $HgSO_4$. The tubes are heated to hasten the oxidation and, as a result, times are shorter. The COD is determined spectrophotometrically. Several commercial designs based on this method are available in the form of an apparatus or kit with solution ampoules and pre-measured reagents.

Total organic carbon

Total organic carbon (TOC) is a method based on the combustion of organic materials in the wastewater sample to CO_2 and water, dehydration of the combustion gases, and running the gases through an infrared analyzer. The analyzer reads out the amount of CO_2 from the combustion, which is proportional to the amount of carbon in the wastewater sample. Sometimes, the presence of inorganic carbon compounds in the wastewater, such as carbonates and bicarbonates, may distort TOC readings, but this problem may be eliminated by purging of inert gases.

Commercial TOC devices employ a different strategy, having two combustion tubes to accommodate combustions of inorganic carbon compounds at 150°C and organic carbon compounds at 950°C. The necessary use of a furnace in the TOC analysis renders this method more expensive, thus preventing TOC analysis from being widely used.

1.1.4 Nitrogen and phosphorous

The sources of nitrogen (N) and phosphorous (P) in food and agricultural wastewater may include chemical fertilizers, synthetic detergents used in cleaning food processing equipment, and metabolic compounds from proteinaceous materials. These elements are nutrients for microbial flora but, if they are present in excess, they may cause proliferation of algae in the receiving water body, with an adverse effect on the ecological balance. Increasingly, many wastewater treatment plants employ advanced wastewater treatment technologies to reduce or eliminate the amounts of nitrogen and phosphorous in the discharge.

1.1.5 Sampling

Accurate characterization of food and agricultural wastewater depends on accurate sampling of wastewater. Special attention should be paid to the representative sampling of a wastewater stream. Commercial sampling instruments are widely available, and simple in-house lab-scale continuous sampler can be set up with relatively modest means (Metcalf & Eddy, Inc., 2002). The procedure for a particular parameter of wastewater management may be found in the standard methods (Eaton et al., 2005).

1.2 Material balances and stoichiometry

In dealing with food and agricultural wastewater, whether formulating treatment and utilization strategy or planning the initial stage of a comprehensive management project, it is essential to have a basic understanding of the effects of mass flow rate or loading factors on process designs.

Stoichiometry is the material accounting for a chemical reaction. Given enough information, one can use stoichiometry to calculate masses, moles, and percents within a chemical equation that is an expression of a chemical process. Consider a simple reaction, where a reactant A converts into resultant B:

$$aA \rightarrow bB \qquad (1.1)$$

where a and b are termed as stoichiometric coefficients and are thus positive proportionality constants.

Equation (1.1) tells us that for every a moles of reactant A consumed, there will be b moles of resultant B produced. If, initially, A has a mole concentration of N_{A0} and B has a starting concentration of N_{B0} then, at any given time, the reactant A and resultant B will be N_A and N_B. They are related to each other by the following expression:

$$(N_{A0}-N_A)/a = (N_B-N_{B0})/b \qquad (1.2)$$

In this expression, $(N_{A0}-N_A)$ represents the consumption of A in moles at the time, while (N_B-N_{B0}) accounts for the gain of B in moles. Equation (1.2) may be used to calculate N_A or N_B when other terms in Equation (1.2) are known. For a more general chemical reaction with the following form:

$$aA + bB \rightarrow cC + dD \qquad (1.3)$$

there will be

$$(N_{A0}-N_A)/a = (N_{B0}-N_B)/b = (N_C-N_{C0})/c = (N_D-N_{D0})/d \qquad (1.4)$$

All terms in the equation are in moles.

Stoichiometric equations stipulate the important principle of mass conservation. Mass can neither be created nor be destroyed; it can only transform from one form or state to another. However, a stoichiometric expression can only provide a snapshot of the underlying chemical reaction at a given time; it does not reveal how fast the chemical reaction goes. For that attribute, we introduce a new term called "chemical reaction rate". Consider the chemical reaction we used in Equation (1.1):

$$aA \rightarrow bB \qquad (1.1)$$

1.2 MATERIAL BALANCES AND STOICHIOMETRY

In this case, we denote the rate of consumption of A per unit volume (molar unit) in a reactor as r_A and the rate of generation of B per unit volume in the reactor as r_B. We know by intuition and the stoichiometric equation:

$$br_A = ar_B \quad (1.5)$$

It should be emphasized that all units discussed so far are mole-based. However, in many biological wastewater treatment process designs and calculations, the units are most likely mass-based. The relationship between mass based units and mole-based units is:

$$[\text{mole} - \text{based units}] = [\text{mass} - \text{based units}]/[\text{molecular weight}] \quad (1.6)$$

It is, however, difficult to establish the exact molecular structures of all microorganisms involved in a wastewater treatment process; therefore, mass-based units have to be used. In this scenario, stoichiometric equations cannot be used, and the relationship between reaction rates needs to be obtained from experiments.

Stoichiometry is a specific form of material balance for reactions and is expressed in mole-based units. In real-world situations, those reactions take place in reactors or other forms of containers. Their designs and layouts will affect the amounts of materials consumed and new substances generated in the reactions. Mass balance equations are used to describe macroscopically the dynamics of materials in a treatment system. We usually start developing mass balance equations on the treatment system with a control volume – a representative portion of the real system that can be integrated over the entire domain of the system. The changes of materials in the control volume should satisfy the law of mass conservation, i.e.:

$$[\text{species in}] - [\text{species out}] + [\text{generation}] = [\text{species accumulation}] \quad (1.7)$$

In mass units, Equation (1.7) can be expressed mathematically as:

$$m_{in} - m_{out} + r_A V_c = d(CV_c)/dt \quad (1.8)$$

Where:
- m_{in} is the mass flow rate of species entering the control volume
- m_{out} is the mass flow rate of species exiting the volume
- V_c is the control volume
- C is the mass concentration of the species.

With appropriate boundary conditions of the system, fluid flow characteristics and the initial condition of the species, Equation (1.8) can be integrated over these conditions to yield the quantities of the variables in the equation.

Equation (1.8) depicts an unsteady state system, where the amount of the species varies with the reaction time. For a steady state system, Equation (1.8) is reduced to:

$$m_{in} - m_{out} + r_A V_c = 0 \qquad (1.9)$$

1.3 Fluid flow rate and mass loading

Almost all wastewater treatment plants are designed based on the annual average daily flow rate of wastewater being processed. However, it should be noted that every plant has to take into account the actual daily flow rate, characteristics of wastewater, and the combination of flow rate and composition (called mass-loading) of the wastewater steam. In an on-site wastewater treatment facility that deals with wastewater streams from a fixed food processing operation, flow rate and mass-loading are not complicated issues in designing and managing wastewater. However, for the wastewater streams from various sources subject to changes in flow rate and mass loadings, peak conditions (whether it is peak flow rate or mass loading) need to be considered as well.

1.4 Kinetics and reaction rates

Chemical or biochemical kinetics is the study of chemical or biochemical reactions with respect to reaction rates, the effect of conditions that reactions are subject to, re-arrangement of molecules, formation of intermediates, and involvement of catalyst. The word "kinetics," originates from the Greek *kinesis*, meaning movement. Thus, kinetics of chemical or biochemical reactions mainly concern the rate of reaction and anything else affecting it.

In general, the reaction rate depends on the concentration of reactants. It may also depend on the concentrations of other species that do not appear in the stoichiometric equation. The dependence of reaction rate on concentrations of reactants can be expressed mathematically in terms of reaction rate constant and the powers of concentrations of reactants. For a general reaction form:

$$aA + bB \rightarrow cC + dD \qquad (1.3)$$

The rate of reaction can be expressed as:

$$r = kC_A^a C_B^b \qquad (1.10)$$

1.4 KINETICS AND REACTION RATES

where:
> k is the reaction rate constant
> a and b are exponents that may or may not be equal to those coefficients appearing in Equation (1.3)
> C_A and C_B are concentrations of reactants A and B.

The sum of a and b is called reaction order, i.e., reaction order for the reaction shown in Equation (1.3) is $(a + b)$. Generally, reactions are categorized as zero-order, first-order, second-order, or mixed-order (higher-order) reactions, based on the value of $(a + b)$. The unit of k is $(concentration)^{1-a-b} (time)^{-1}$.

1.4.1 Zero-order reactions

Zero-order reactions (order = 0) have a constant rate. This rate is independent of the concentration of the reactants. The rate law is:

$$r = k$$

with k having the units of $(concentration)^1 (time)^{-1}$, e.g., M/sec.

1.4.2 First-order reactions

A first-order reaction (order = 1) has a rate proportional to the concentration of one of the reactants. A common example of a first-order reaction is the phenomenon of radioactive decay. The rate law is:

$r = kC_A$ (or C_B instead of C_A), with k having the units of $(time)^{-1}$, e.g., sec^{-1}.

1.4.3 Second-order reactions

A second-order reaction (order = 2) has a rate proportional to the concentration of the square of a single reactant or the product of the concentration of two reactants:

rate = kC_A^2 (or substitute B for A or k multiplied by the concentration of A times the concentration of B), with the unit of the rate constant
$k = (concentration)^{-1}(time)^{-1}$, e.g., $M^{-1} sec^{-1}$

1.4.4 Mixed-order or higher-order reactions

Mixed-order reactions, such as some biochemical reactions, have a fractional order for their rate. e.g.:

$$\text{rate} = kC_A^{1/3}$$

The unit of the rate constant k is (concentration)$^{2/3}$ (time)$^{-1}$, e.g., M$^{2/3}$/sec.

1.4.5 Catalytic reactions

Almost all biochemical reactions involve catalysts – enzymes that are specialized proteins synthesized by microorganisms. A catalyst is a substance (enzyme for a biocatalyst) that increases the rate of reaction without undergoing *permanent* (bio)chemical change. The primary function of a catalyst is to lower the activation energy of a reaction, so that the reaction can be carried out easily, but not to affect the reaction equilibrium. In biochemical reactions, the enzyme is believed to possess certain active sites, consisting of amino acid side chains or functional groups, to which the specific functional groups of substrate molecules bind. Thus, the enzyme is reaction-specific. The active sites of the enzyme act as the donors or acceptors of electrons from the substrate molecules, and speed up the reaction. It is assumed that the enzymatic reaction involves a series of step-by-step elementary reactions forming complexes with substrate molecules along the way. It is described by Michaelis-Menten kinetics:

$$E + S \underset{k_{-1}}{\overset{k_1}{\rightleftharpoons}} ES \overset{k_2}{\longrightarrow} E + P \qquad (1.11)$$

The terms k_1, k_{-1} and k_2 are rate constants for, respectively, the association of substrate and enzyme, the dissociation of unaltered substrate from the enzyme and the dissociation of product (= altered substrate) from the enzyme. The overall rate of the reaction (r_P) is limited by the step ES to E + P, and this will depend on two factors: the rate of that step (i.e., k_2) and the concentration of enzyme that has substrate bound, i.e., C_{ES}:

$$r_P = k_2 C_{ES} \qquad (1.12)$$

At this point, we make two assumptions. The first is the availability of a vast excess of substrate, so that $C_S \gg C_E$. The second is that it is assumed that the system is in pseudo-steady state, i.e., that the ES complex is being formed and broken down at the same rate, so that overall C_{ES} is constant. The formation of ES will depend on the rate constant k_1 and the availability of enzyme and

1.4 KINETICS AND REACTION RATES

substrate, i.e., C_E and C_S. The breakdown of C_{ES} can occur in two ways – either the conversion of substrate to product or the non-reactive dissociation of substrate from the complex. In both instances, the C_{ES} will be significant. Thus, at steady state, we can write:

$$k_1 C_E C_S = (k_{-1} + k_2) C_{ES} \tag{1.13}$$

The term,

$$(k_{-1} + k_2)/k_1 = K_m \tag{1.14}$$

where K_m is called the Michaelis-Menten constant.

The total amount of enzyme in the system must be the same throughout the experiment, but it may either be free (unbound) E or in complex with substrate, C_{ES}. If we term the total enzyme C_{E0}, this relationship is expressed as:

$$C_{E0} = C_E + C_{ES} \tag{1.15}$$

in which C_{E0} represents initial enzyme concentration.

Inserting Equations (1.15) and (1.14) into Equation (1.13) and re-arranging the resulting equation lead to:

$$C_{ES} = C_{E0} C_S / (K_m + C_S) \tag{1.16}$$

So, substituting this right-hand side into Equation (1.12) in place of C_{ES} results in:

$$rp = k_2 C_{E0} C_S / (K_m + C_S) \tag{1.17}$$

The maximum rate, which we can call r_{max}, would be achieved when all active sites of the enzyme molecules have saturated with substrate molecules. Under conditions when C_S is much greater than C_E, it is reasonable to assume that all C_E will be in the form C_{ES}. Therefore, $C_{E0} = C_{ES}$. We may substitute the term r_{max} for r and C_{E0} for C_{ES} in Equation (1.12), which would give us:

$$r_{max} = k_2 C_{E0} \tag{1.18}$$

So, we now have:

$$rp = r_{max} C_S / (K_m + C_S) \tag{1.19}$$

This equation is commonly referred to as Michaelis-Menten equation.

The significance of Michaelis-Menten equation is that when r_p is half of r_{max}, from Equation (1.19), we would have

$$C_S = K_m \tag{1.20}$$

The K_m of the enzyme is the substrate concentration at which the reaction occurs at half of the maximum rate and is, therefore, an indicator of the affinity that the enzyme has for a given substrate and, hence, the stability of the enzyme-substrate complex. This interpretation may be better presented by plotting r_p versus C_S, which is called the Michaelis plot, shown in Figure 1.2.

It is obvious that, at low C_S, it is the availability of substrate that is the limiting factor. Therefore, as more substrate is added, there is a rapid increase in the initial rate of the reaction – any substrate is rapidly gobbled up and converted to product. At the K_m, 50% of active sites have substrate occupied. At higher C_S, a point is reached (at least theoretically) where all sites of the enzyme have substrate occupied. Adding more substrate will not increase the rate of the reaction; hence, the leveling-out observed in Figure 1.2.

In order to use the Michaelis-Menten equation, one needs to know the values of K_m and r_{max}. The common approach is to linearize the Michaelis-Menten equation by plotting $1/r_p$ versus $1/C_S$ (called Lineweaver-Burk linearization), which results in a slope of the linearized line, K_m/r_{max} and an intercept on the $1/r_p$ axis, $1/r_{max}$. Other linearization schemes of the Michaelis-Menten equation, such as Hanes-Wolf and Eadie-Hofstee plots, would accomplish the same objective as Lineweaver-Burk linearization.

1.5 Theoretical modeling and design of biological reactors

Theoretical modeling of biological wastewater reactors using mathematical equations allows engineers and designers to test their strategies and to evaluate their treatment options virtually, therefore reducing the amounts of time and money as well as the potential hazardous incidents that could happen to an actual experimentation. In an existing system, a robust model can be used to optimize the operational strategies. The development of the model often involves the selection of suitable equations that accurately describe fluid flow in the reactor and biochemical reactions in the form of microbial growth on organic and inorganic materials in the reactor. Many equations derived hereafter are more or less simplified equations of the generic reactor types.

This approach has its own advantages. First, it acknowledges that most biological reactors in use for wastewater treatment are quite similar to the generic reactors described below. Second, the methodologies of derivation of the equations for the generic reactors are valid for more "realistic" or complex reactors. Some of those equations related to reaction kinetics, mass balance, stoichiometry, and chemical thermodynamics have been explained previously. The overriding goal of this section is to combine fluid flow with kinetics in several geometrical

1.5 THEORETICAL MODELING AND DESIGN OF BIOLOGICAL REACTORS

environments of the generic reactor types in order to derive the reaction rate expressions and concentration profiles of substrates in the reactors. For the sake of simplicity, we focus on our attention initially to single reactions occurring in the liquid phase of constant density in single reactors.

1.5.1 Batch reactors

In a batch reactor, at any given time since the reactor starts, there is feed neither coming in nor coming out. The mass balance of a batch reactor from Equation (1.8) will be:

$$0 - 0 + r_A V_c = d(CV_c)/dt \tag{1.21}$$

For a constant volume, the above equation is:

$$r_A = dC_A/dt \tag{1.22}$$

This may be integrated from the initial concentration of A, C_{A0} to the final concentration C_{Af}, i.e.:

$$\int dt = t = \int dC_A/r_A \tag{1.23}$$

The exact relationship between r_A and C_A (kinetics) needs to be known in order to solve Equation (1.22) and establish the concentration history of reactant A.

For zero-order reactions (order = 0), r = k, so Equation (1.22) develops into:

$$t = \int dC_A/r_A = \int dC_A/k = (C_{A0} - C_{Af})/k \tag{1.24}$$

For first-order reactions (order = 1), $-r = kC_A$, so Equation (1.22) becomes:

$$t = \int dC_A/r_A = -\int dC_A/kC_A = \ln(C_{A0}/C_{Af})/k \tag{1.25}$$

For second-order reactions (order = 2), $-r = k(C_{B0} - C_{A0} + C_A)C_A$, so Equation (1.22) turns into:

$$C_{Af}/C_{A0} = (C_{Bf}/C_{B0}) \exp[-(C_{B0} - C_{A0}) kt] \tag{1.26}$$

where:
$C_{Bf} = C_{Af} - C_{A0} + C_{B0}$
C_{B0} is the initial concentration of reactant B.

If $C_{A0} = C_{B0}$, $-r = kC_A^2$ and Equation (1.22) will yield

$$kt = 1/C_{Af} - 1/C_{A0} \qquad (1.27)$$

1.5.2 Continuous stirred tank reactors (CSTRs)

Continuous stirred tank reactors (CSTRs) are widely used in biological wastewater treatment processes and can be schematically viewed as tanks with input and output, while a mechanical or pneumatic device provides the means of thorough mixing the liquid phase in tanks. In CSTRs, the liquid inside the reactor is completely mixed. The mixing is provided through an impeller, rising gas bubbles (usually oxygen) or both. The most characteristic feature of a CSTR is that it is assumed that the mixing is uniform and complete, such that the concentrations in any phase do not change with position within the reactor.

The dissolved oxygen in the tank is the same throughout the bulk liquid phase. Because of this uniformity of oxygen distribution in the reactor, a CSTR for wastewater treatment operations has the advantage of de-coupling the aerator or stirrer from the reaction as long as oxygen is well provided for (no need to consider pesky fluid mechanics), thus simplifying process design and optimization. Under the steady state, where all concentrations within the reactor are independent of time, we can apply the following materials balance on the reactor:

$$\begin{bmatrix} \text{Rate of addition} \\ \text{to reactor} \end{bmatrix} + \begin{bmatrix} \text{Rate of accumulation} \\ \text{within reactor} \end{bmatrix} = \begin{bmatrix} \text{Rate of removal} \\ \text{from reactor} \end{bmatrix} \qquad (1.28)$$

Replacing the statements in the above expression with mathematical symbols leads to:

$$FC_{A0} + V_R r_A = FC_A \qquad F(C_{A0} - C_A) = -V_R r_A \qquad (1.29)$$

where F is volumetric flow rate of feed and effluent liquid streams.

Re-arrangement of Equation (1.29) yields:

$$r_A = \frac{F}{V_R}(C_A - C_{A0}) = D(C_A - C_{A0}) \qquad (1.30)$$

where $D = F/V_R$ and is called the dilution rate. The term characterizes the holding time or processing rate of the reactor under steady state condition. It is the number of full-tank volumes passing through the reactor tank per unit time and is equal to the reciprocal of the mean holding time of the reactor.

Because of lack of time dependence of concentrations in CSTR and, thus, the differential form of reactor analysis as in a batch reactor, CSTRs have the advantage of being well-defined, easily reproducible reactors. They are used frequently in many cell growth kinetics studies, despite relatively high cost and a long time for achieving steady state. Batch reactors, which can be as simple as sealed beakers or flasks used in an incubator shaker, are still widely used for their inexpensive, fast, and unbridled benefits.

No matter what type of reactors are used, the goal of studying cell growth kinetics should be based on the intended application and scope of the use of the kinetics. Only then may the experimental design and implementation be formulated.

1.6 Process economics

Process economics is the next step of a wastewater treatment and management design project, after preliminary selections of wastewater treatment processes have been completed in accordance with the project objectives. The economical considerations of the wastewater treatment and management project, including aspects of material and energy usage and recovery, are among the most important factors that influence the final decision about the project.

To develop meaningful cost estimates, the data from the wastewater characterization and other possible alternatives to the selected processes should be available. The cost estimates of the unit operations in wastewater treatment and management operations can be evaluated with the cost correlations developed by US EPA (1983). The cost correlations for alternative processes should also be gathered prior to the final estimation.

1.6.1 Capital costs

Capital costs usually refer to the process unit construction costs, the land costs, the costs of treatment equipment, financial costs in association with loan and services, costs of environmental impact or other community-imposed costs, and the costs of engineering, administration and contingencies.

1.6.2 Operational costs and facility maintenance

There are several important factors that determine the operational costs: energy costs, labor costs, materials and chemical costs, costs of transportation of treated sludge and treated wastewater, and discharge costs. The relative importance of these costs is highly dependent on locality and the quality of the influent and effluent of the wastewater treatment plant.

1.7 Further Reading

Middlebrooks, E.J. (1979a). *Industrial Pollution Control, Vol. 1: Agro-Industries.* John Wiley & Sons, New York, NY.

1.8 References

Bauman, F.J. (1974). Dichromat reflux chemical oxygen demand. A proposed method for chlorine correction in highly saline wastes. *Analytical Chemistry* **46**, 1336–1338.

Eaton, A.D., Clesceri, L.S., Rice, E.W., Greenberg, A.E. & Franson, M.A.H. (2005). *Standard Methods for Examination of Water & Wastewater: Centennial Edition (Standard Methods for the Examination of Water and Wastewater).* American Public Health Association, Washington, D.C.

Liu, S.X., Hermanowicz, S.W. & Peng, M. (2003). Nitrate removal from drinking water through the use of encapsulated microorganisms in alginate beads. *Environmental Technology* **22**, 1129–1134.

Magbunua, B. (2000). *An Assessment of the recovery and Potential of Residuals and By-Products from the Food Processing and Institutional Food Sectors in Georgia.* University of Georgia Engineering Outreach Services, Athens, GA.

Middlebrooks, E.J. (1979b). *Industrial Water Pollution Control.* John Wiley & Sons, New York, NY.

Metcalf & Eddy, Inc. (2002). *Wastewater Engineering: Treatment and Reuse,* 4th edition. McGraw-Hill Science/Engineering/Math, New York, NY.

Peng, M., Vane, L.M. & Liu, S.X. (2003). Recent Advances in VOC Removal from Water by Pervaporation. *Journal of Hazardous Materials* **B98**, 69–90.

Soderquist, M.R., Blanton, G.I. & Taylor, D.W. (1975). *Characterization of Fruit and Vegetable Processing Wastewaters.* Water Resources Research Institute, Oregon Agricultural Experiment Station Technical Paper No. 3388, Oregon State University.

US Environmental Protection Agency (1974). *Development Document for Effluent Limitations Guidelines and New Source Performance Standards for the Red Meat Processing Segment of the Meat Product and Rendering Processing Point Source Category.* US EPA, Washington D.C.

US Environmental Protection Agency (1983). *Construction Costs For Municipal Wastewater Treatment Plants: 1973–1982.* EPA Publication 430983004. US EPA, Washington D.C.

2
Basic microbiology in wastewater treatment

2.1 Introduction

Microorganisms have several critical functions in wastewater treatment processes. It is the microbial component of the aquatic ecosystem that provides the purifying capacity of natural waters in which microorganisms respond to the rise of organic pollutant concentration by increased growth and metabolism. The self-regulated purification mechanism in natural waters operates on the same fundamental principles of biological pollutant abatement as those used in biological wastewater treatment processes. In addition to having abundant food for microorganisms, food and agricultural wastewaters also contain microorganisms themselves. With a controlled and suitable environment for optimal growth and metabolism of microorganisms, all the organic matter extant in the wastewater streams can be biodegraded.

Microorganisms utilize organic materials for the production of energy by cellular respiration, and also for the synthesis of cellar material such as proteins for maintenance and production of new cells. The utilization of organic matters by microorganisms can be summarized in an equation showing the overall reaction of wastewater treatment:

$$\text{Organic matter} + O_2 + NH_4^+ + P \rightarrow \text{New cells} + CO_2 + H_2O \quad (2.1)$$

where P stands for phosphate.

In wastewater treatment, bacteria are the organisms responsible for the degradation of organic matter. Other microorganisms, such as fungi, algae, protozoa, and higher organisms, also contribute to the biotransformation of organic materials in wastewaters.

A basic understanding of the key biological organisms, microbial or otherwise – bacteria, fungi, algae, protozoa, and metazoan (crustaceans) – is essential in developing strategies for food and agricultural wastewater treatment

Food and Agricultural Wastewater Utilization and Treatment, Second Edition. Sean X. Liu.
© 2014 John Wiley & Sons, Ltd. Published 2014 by John Wiley & Sons, Ltd.

and utilization. With the exception of wastewater streams coming directly from certain food processing operations such as retort, food and agricultural wastewater provides an ideal growth medium where diverse communities of microorganisms can thrive. These organisms play an important role in all stages of biological wastewater treatment, and they also have some influences on sludge formation and characteristics. It is no exaggeration to say that a biological wastewater treatment process could not survive if were not for the existence of mixed communities of ravenous microorganisms feeding on organic matter and each other. The exact compositions of these communities will depend on the outcome of competition of food supplies among microbial species and environmental conditions such pH and temperature. Information pertaining to microbial communities at each stage of biological wastewater treatment is thus important for designers and planners of food and agricultural wastewater management.

Although, as we are taught from our elementary school days, living things are classified as either plant kingdom or animal kingdom, microorganisms belong to a unique kingdom called "Protista", due to the fact that mixed communities of mostly unicellular microorganisms have plant-like constituents such as unicellular algae, animal-like inhabitants such as protozoa, and fungus-like creatures such as water molds. These are simple life forms whose cells possess various structures.

2.2 Structures of cells

Typical structures of cells include prokaryotes (which encompass all bacteria) and eukaryotes (which are the building blocks of all living organisms except bacteria and viruses). Bacterial cells are molecules surrounded by a semi-permeable membrane which, in turn, is covered by a porous cell wall. The protoplasm in the membrane contains nuclear material that is not bounded by any other material separating it from the protoplasm. Eukaryotes are cells that contain complex and well-defined internal structures (organelles) surrounded by membranes. Characteristic organelles comprise the cell nucleus, the mitochondria, and the ribosomes. Figure 2.1 shows a diagram of a cell structure of a bacterium.

2.3 Important microorganisms in wastewater

2.3.1 Bacteria and fungi

Bacteria are the most important and the largest components of the microbial community in all biological wastewater treatment processes. Depending on the

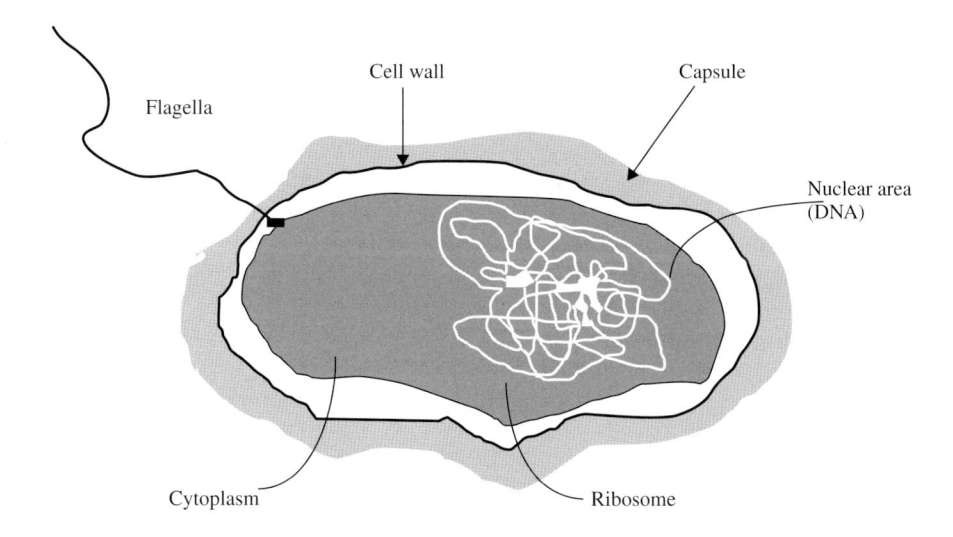

Figure 2.1 A schematic diagram of a bacterium cell

biological process and the pH, the number of bacteria present will vary, with activated sludge (aggregates of healthy aerobic flocs) having the largest number concentration of bacteria. Bacteria range in size from approximately 0.5 to 5 μm and take one of four major shapes: sphere (cocci), straight rod, curved rod (vibrio), and spiral (spirilla). They appear in singly, in pairs, packets, or in chains. Bacteria are classified into two major groups: heterotroph (that use organic matters as both energy and carbon sources for synthesis) and autotroph (that use inorganic matters for energy source and CO_2 for carbon source). The heterotrophs can be further subdivided into three categories: aerobic (using free oxygen for decomposing organic matters), anaerobic (using no free oxygen for decomposing organic matters), and facultative (thriving both in aerobic and in anaerobic environments).

Aerobic bacteria require free dissolved oxygen to decompose organic materials:

$$\text{Organics} + O_2 \xrightarrow{\text{aerobes}} \text{Aerobes} + CO_2 + H_2O + \text{energy} \quad (2.2)$$

This microbial reaction is autocatalytic, meaning that the bacteria that function as catalysts are also produced by the reaction. Aerobic bacteria are predominant in biological wastewater treatment processes such as activated sludge and trickling filters and other biological processes that utilize free oxygen for their biochemistry. Aerobic bioconversion of organic matters is a biochemically efficient and rapid process that results in various products with highly oxidized compounds such as CO_2 and water.

The metabolism of aerobic bacteria is much higher than that of anaerobic bacteria. This augmentation means that 90% fewer organisms for aerobic metabolism are needed compared to the anaerobic process, and treatment is accomplished in 90% less time. This provides a number of advantages, including a higher percentage of organic removal. Aerobic bacteria live in colonial structures called floc and are kept in suspension by the mechanical action used to introduce oxygen into the wastewater. This action exposes the floc to the organic matter while biological treatment takes place. Following digestion, a gravity clarifier separates and settles out the floc.

Anaerobic bacteria live and reproduce in the absence of free oxygen. They utilize compounds such as sulfates and nitrates for energy and their metabolism is substantially reduced. In order to remove a given amount of organic matters in an anaerobic environment, the organic materials must be exposed to a significantly higher quantity of bacteria and/or engaged for a much longer period of time. A representative use for anaerobic bacteria would be in a septic tank.

The slower metabolism of anaerobic bacteria requires that the wastewater be held for several days in order to achieve even a nominal 50% reduction in organic matters. However, the advantage of using the anaerobic process is that mechanical equipment is not required. Anaerobic bacteria release hydrogen sulfide as well as methane gas, both of which can create hazardous conditions for humans and animals. The following reactions represent the anaerobic transformation by anaerobes common in wastewater treatment:

$$\text{Organics} + NO_3 \xrightarrow{\text{anaerobes}} \text{Anaerobes} + CO_2 + N_2 + \text{energy} \quad (2.3)$$

which utilizes bounded oxygen in nitrate; or

$$\text{Organics} + SO_4^{2-} \xrightarrow{\text{anaerobes}} \text{Anaerobes} + CO_2 + H_2S + \text{energy} \quad (2.4)$$

which utilizes bounded oxygen in sulfate.

Anaerobic bacterial activities are found primarily in the digestion of sludge and wastewater lagoons. Anaerobic processes are normally biochemically inefficient and generally slow, producing complex end products that often emit obnoxious smells. In food and agricultural wastewater treatment, in addition to being degraded into amino acids and CO_2 (like aerobic degradation), proteins are often also degraded anaerobically into hydrogen, alcohols, organic acids, methane, hydrogen sulfide, phenol, and indol.

Most of the bacteria that absorb organic matter in a wastewater treatment system are facultative in nature, and the nature of individual facultative bacteria is dependent on the environment in which they live. Usually, facultative bacteria such as *E. coli* will be anaerobic, unless there is some type of mechanical or

2.3 IMPORTANT MICROORGANISMS IN WASTEWATER

Table 2.1 Common organisms encountered in biological wastewater treatment (excluding flies)

Species	Genre	Process involved
Achromobacter	Bacteria	Biofilters and activated sludge
Acinetobacter	Bacteria	Biological phosphorous removal
Alcaligenes	Bacteria	Biofilters, activated sludge, and sludge digester
Bloodworm	metazoa	Biofilters and treated sludge
Chironomus	metazoa	Stabilization ponds and sludge
Crustacea	metazoa	Stabilization ponds and activated sludge
Daphnia	metazoa	Activated sludge and ponds
Desulfovibrio	Bacteria	Sludge digesters
Flavobacterium	Bacteria	Activated sludge, biofilters, sludge digester
GAO	Bacteria	Biological phosphorus removal
Geotrichum	Fungus	Activated sludge and biofilters
Gordonia	Bacteria	Activated sludge
Micrococcus	Bacteria	Activated sludge and biofilters
Microtrix	Bacteria	Activated sludge
Nitrobacter	Bacteria	Nitrification
Nitrosomonas	Bacteria	Nitrification
PAO	Bacteria	Biological phosphorus removal
Pseudomonas	Bacteria	Denitrification
Rotifera	Metazoa	Activated sludge
Sphaerotilus natans	Bacteria	Activated sludge
Tubifex	Metazoa	Biofilters
Vorticella	Protozoa	All aerobic processes and ponds
Zoogloea ramigera	Bacteria	Activated sludge and biofilters

biochemical process used to add oxygen to the wastewater. When bacteria are in the process of being transferred from one environment to the other, the metamorphosis from anaerobic to aerobic state (and vice versa) takes place within a couple of hours. Common bacteria found in biological wastewater treatment processes are listed in Table 2.1.

Using glucose as the organic substance and the formula $C_5H_7O_2N$ to represent the composition of microorganisms, the basic organic bioconversion brought about by aerobes in biological wastewater treatment plants may be represented by the following equation:

$$C_6H_{12}O_6 + 0.5NH_4^+ \rightarrow C_5H_7O_2N + 3.5CO_2 + 5H_2O + 0.5H^+ \quad (2.5)$$

Many studies on cell compositions have revealed that bacteria are comprised of 80% water and 20% of dry matter, with approximately 90% of the dry matter in bacteria being organic. An approximate formula of $C_5H_7O_2N$ is often used in expressing a biochemical reaction, but the formulation $C_{60}H_{87}O_{23}N_{12}P$ may also be used when phosphorus is considered. The inorganic compounds in cells are about 50% phosphorus, 15% sulfur, 11% sodium, 9% calcium, 8% magnesium, 6% potassium, and 1% iron. All of these inorganic elements are required for microbial growth and, since all of them are derived from the environment, a shortage of any of these elements would result in stunted growth or an altered growth path. The pH of the environment is also important in microbial activities. Most bacteria cannot tolerate pH levels above 9.5 or below 4.0. The optimum pH value range of optimal growth for bacteria lies between 6.5 and 7.5.

Fungi are a group of microscopic non-photosynthetic plants including yeasts and molds. Yeasts are widely used in the food industry for brewing and baking, and molds are filamentous fungi that bear a resemblance to higher plants in structure, with branched, fractal-like growths. Fungi tend to compete disadvantageously with bacteria for nutrients; their numbers are low except when pH is low, because acidic conditions favor the growth of fungi. Fungi such as molds are nuisances in many biological wastewater processes, because of their filamentous nature, which interferes with floc settling in flocculation and sedimentation basins.

The majority of filamentous organisms are bacteria, although some of them are classified as algae, fungi or other life forms. There are a number of types of filamentous bacteria that proliferate in the activated sludge process. Filamentous organisms perform several different roles in the process, some of which are beneficial and some of which are detrimental.

When filamentous organisms are in low concentrations in the process, they serve to strengthen the floc particles. This effect reduces the amount of shearing in the mechanical action of the aeration tank and allows the floc particles to increase in size. Larger floc particles settle more readily in a clarifier. Larger floc particles settling in the clarifier also tend to accumulate smaller particulates (surface adsorption) as they settle, producing an even higher quality effluent.

Conversely, however, if the filamentous organisms reach too high a concentration, they can extend dramatically from the floc particles and tie one floc particle to another (inter-floc bridging), or even form a filamentous mat of extra large size. Due to the increased surface area without a corresponding increase in mass, the activated sludge will not settle well. This results in less solid separation and may cause a washout of solid material from the system. In addition, air bubbles can become trapped in the mat and cause it to float, resulting in a floating scum mat. Due to the high surface area of the filamentous bacteria, once they reach an excess concentration, they can absorb a higher percentage of the organic matter and inhibit the growth of more desirable organisms.

2.3.2 Algae

Algae are photosynthetic eukaryotes that inhabit in all water bodies. There are only two situations where algae become involved in wastewater: trickling filters and stabilization bonds. Only stabilization ponds utilize algae to treat wastewater. The distinct feature of algae is that they use photosynthesis to produce energy via chlorophyll (the source of the green color in green plants). The major group of algae is the green algae found in aquatic environments. Blue-green algae are prokaryotes.

2.3.3 Protozoa and metazoa

In a wastewater treatment system, the next higher life form above bacteria is protozoa. These single-celled animals perform three significant roles in the activated sludge process: floc formation; cropping of bacteria; and the removal of suspended material. Protozoa are also indicators of biomass health and effluent quality.

Because protozoa are much larger in size than individual bacteria, identification and characterization is readily performed. Four major groups of protozoa have been identified: Mastigophora, Sarcodina, Sporazoz, and Ciliata. Ciliates are the largest and most important protozoa in biological wastewater treatment, where they feed on bacteria and aid in both bioflocculation and clarification.

Metazoans are very similar to protozoa, except that they are usually multi-celled animals. Macro-invertebrates such as nematodes and rotifers are typically found only in a well-developed biomass. The presence of protozoa and metazoans, and the relative abundance of certain species, can be a predictor of operational changes within a wastewater treatment plant. In this way, a wastewater treatment plant operator is able to make adjustments based on observations of changes in the protozoan and metazoan population, in order to minimize negative operational effects.

2.3.4 Role of microorganisms in biological wastewater treatment

The role of microorganisms in wastewater treatment varies with the specific biological process and the environment the microorganisms are in. In the activated sludge process, which is operated often as a BOD reducer, the flocs that characterize the essence of the activated sludge process is comprised of microorganisms, organic matters, inorganic colloidal materials and larger particulates.

The structures of flocs provide certain advantages, serving not only as colonies for BOD removal agents such as bacteria, but also as traps for soluble and insoluble BOD, where they are readily hydrolyzed by extracellular enzymes prior to being absorbed and metabolized by microorganisms. Another important function for activated sludge is its significant role in promoting good settlement in the secondary sedimentation tanks or basins.

In trickling filter, the role of microorganisms in wastewater treatment is played out in the slime layer (called the biofilm) that adheres to the surface of the supporting media (also known as the filter media). Trickling filter is a biological wastewater treatment system that consists of a circular bed of coarse stones and plastics that are continuously subject to a trickling flow of wastewater from an overhead rotating distributor. The bacteria in the wastewater attach themselves to the bedding materials, where the organic materials break down.

Slime-producing bacteria such as *Z. ramigera* often initiate the formation of, and the thickening of, the biofilm. However, many other organisms contribute to further colonization of the biofilm as a multi-layered structure with an outer layer often comprised of fungi and removal of BOD from the passing wastewater. At least in terms of composition of the biomass of the biofilm, this is in contrast to that of activated sludge, where the existence and role of algae and fungi are insignificant. The design of trickling filters also may limit mass transfers of O_2, soluble organic matters and metabolized substances. This is more a severe issue when the thickness of the biofilm is considerable enough to affect the biodegradation of organic matter.

Anaerobic digestion is a slower process of biodegradation of organic matter by anaerobic and facultative bacteria, and it is usually carried out in a continuously stirred tank reactor (CSTR) in order to suspend the insoluble organic materials. The reactor usually has a residence time of several days and is fed with a slurry of solids. The end products from the reactor usually are solids (less than the feed), carbon dioxide and methane.

Two groups of bacteria are involved; one group, comprised of non-methanogenic bacteria, converts organic matters to simpler compounds such as organic acids, carbon dioxide and hydrogen. The second group, called methanogenic bacteria, transforms the metabolized products into methane. The interdependence between these two groups of bacteria fosters a delicate relationship that could be easily out of sync with some changes in general environmental parameters, such as pH. The fragile alliance of the bacteria in these systems contributes to the difficulty in operating anaerobic digesters.

On the other hand, the role of the wastewater stabilization pond, a condensed ecosystem of nature, is more complex, despite its simplistic appearance and operational logistics. It has diverse species of biota and incorporates several complete nutrient recycles: carbon, nitrogen, and sulfur. Depending on the primary purposes of the ponds, they can be divided into three basic groups: anaerobic, facultative, and maturation. The microorganisms inhabiting these communities vary considerably in terms of the dominant species within the ponds. In general, the higher the BOD, the lower diversity of species in these ponds.

2.4 Microbial metabolism

2.4.1 Microbial energy generation

Microorganisms consume energy for their growth, reproduction, and maintenance (tasks such as motility – i.e., transport of materials in and out of the cell – and synthesis of new cell materials). The energy is derived from either a physical source (light) or a chemical source (breakdown of substrates). It is converted into biologically utilizable energy by microbial metabolism and stored inside the microorganism in a chemical form, as a compound called adenosine 5'-triphosphate or ATP for short, which consists of an adenosine molecule that is linked to three inorganic phosphate molecules by phosphoryl bonds. These bonds are the energy source for microbial activities, since their formation requires a large amount of energy, and the hydrolysis of the bonds releases energy that can be utilized microbiologically.

The production of ATP is through the reaction between adenosine 5'-diphosphate (ADP) and inorganic phosphate, resulting in a new phosphoryl bond in the ATP. Once the ATP is formed, it can be stored in the cell and used as needed by hydrolyzing the phosphoryl bond. Two types of phosphorylation reactions form ATP: substrate-level and oxidative.

The substrate-level phosphorylation is particularly important for certain bacterial growth that is devoid of free oxygen. The anaerobic microbes synthesize ATP exclusively, with one of six inorganic phosphate compounds (substrates) in an enzymatic reaction. The other source of the microbial energy generation can be viewed as a biological redox half-reaction of NAD(P)H and NADH, with nicotinamide adenine dinucleotide (NAD^+) and its phosphorylated product of $NADP^+$ as acceptors for electrons. The oxidation of NADH and NAD(P)H releases energy to synthesize ATP. For example, oxidation of one mole of NAD(P)H helps yield three mole of ATP.

2.4.2 Uptake of substrates into microbial cell

Microbial growth requires the substrates in the wastewater to be brought inside the cell for utilization. Not all organic particulates or soluble solids can penetrate through the rigid, hydrophobic cell wall of a bacterium; only small hydrophobic molecules can permeate through the cell membrane unassisted. Some species of microorganisms have the ability to secrete enzymes outside the cell, hydrolyzing the larger molecules into smaller soluble molecules that can enter the cell. Any form of close contacts between substrates and the cells enhances the enzymatic breakdown, as is the case for trickling filters or activated sludge processes.

The small hydrophobic substrate molecules can permeate through the cell wall of a microorganism if the concentration of the substrate molecules is higher than the concentration of the substrate concentrations across the cell wall and inside the cell, via a molecular diffusion mechanism. The mechanism of this diffusional mass transfer is similar to the mass transfer of molecules across a synthetic membrane. The majority of substrate transport, however, relies on a more active form of mass transfer that requires energy. As described previously, ATP contains energy-rich phosphoryl bonds that can be broken by hydrolysis and a large amount of energy released. Hence, ATP hydrolysis is an exergonic reaction. However, the release of energy from hydrolysis is not in the form of heat but, rather, it is used to drive the coupled biological reactions that need energy to complete (these reactions are called endergonic reactions).

A portion of ATP energy is used for the active transport of substrates. This type of active mass transport of substrates requires a group of carrier enzymes called permeases, which combines the word "permeate" and the suffix "-ases" for enzymes. Permeases are substrate-selective and, therefore, the uptakes of substrates in wastewater often are restricted by the amount of permeases present. The permeases-assisted substrate transport overcomes the limit associated with the requirement of concentration gradient across the cell wall for molecular diffusion. It is not unusual to have internal substrate concentration inside the bacterial cell up to a thousand-fold higher than the level in the wastewater. This is also the reason why the microorganisms can live in low-BOD environments, such as rivers and oceans.

2.4.3 Oxidation of organic and inorganic substrates

Organic matter in wastewater is not directly oxidized to CO_2 and H_2O, because there is no energy conservation mechanism accommodating the release of large amounts of energy resulting from oxidations with CO_2 and H_2O as the end products. Rather, they are oxidized in small steps. This typically involves the transfer of an electron from the substrate being oxidized to some acceptor molecule, which will be reduced as a result. In microbial cells, the major electron acceptors (sometimes also called hydrogen acceptors, due to the fact that for every electron removal there is a simultaneous loss of a proton, and the net result of this is the loss of a hydrogen atom) are two carrier molecules known as pyridine nucleotides: NAD and NADP. When they undergo redox reactions, the energy released from oxidation of NAD and NADP helps to synthesize ATP. Microorganisms that obtain their reducing equivalents necessary for energy generation from oxidation of organic matters are called organotrophs (including photoorganotrophs that derived energy from sunlight for photosynthesis, and chemoorganotrophs that generate energy from oxidation of organic compounds).

Many microorganisms are also able to oxidize inorganic materials. These microbes are termed lithotrophs. Bacteria that obtain their energy from the oxidation of inorganic compounds coupled with the energy release to ATP synthesis by means of electron transport chain are called chemolithotrophs. Those lithotrophs that derive their energy directly from sunlight are also known as photolithotrophs. There are many potential inorganic energy sources that include hydrogen, ammonia, metal ions (e.g., Fe^{2+}), and sulfur for biochemical reactions in the wastewater treatment.

2.5 Nitrification

Nitrification is a microbial process that converts ammonia into nitrite and, ultimately, into nitrate. Ammonia in wastewater comes primarily from two sources: intense use of nitrogen-rich fertilizers such as urea; and organic nitrogen from proteins. The de-amination of organic nitrogen and hydrolysis of urea under urease results in ammonia:

$$\text{Urea} + 2H_2O \xrightarrow{\text{urease}} (NH_4^+)_2 CO_3^{2-} \quad (2.6)$$

$$\text{Amino acid} + 0.5 O_2 \rightarrow R - CO - COOH + NH_4 \quad (2.7)$$

for oxidative de-amination, and:

$$\text{Amino acid} + 2H \rightarrow R - CH_2 - COOH + NH_4 \quad (2.8)$$

for reductive de-amination.

Wastewaters from the fishery, meat and poultry industries contain substantial amount of proteins. By the time of these proteins reach the collection facilities of the wastewater treatment plants, the majority of them have been converted into peptides and amino acids by extracellular proteolytic enzymes and, ultimately, into ammonia.

In nitrogen-rich wastewater streams, only small amount of the nitrogen is removed from wastewater through conventional heterotrophic activity by incorporating nitrogen into the microbial biomass. There is a parallel in phosphorus-rich wastewaters. If left untreated, the nitrogen or/and phosphorus in discharged effluents will cause eutrophication (a form of photo-autotrophic activity) of the receiving waters, which will gravely disrupt the aquatic ecosystem.

The nitrification process in biological wastewater treatment (i.e., the use of a limited group of autotrophic nitrifying bacteria to convert ammonia into nitrite and, eventually, nitrate), is often used in the so-called advanced wastewater treatment phase of a wastewater treatment scheme, if the concentration of ammonia in wastewater streams is high enough to warrant the treatment. Nitrification is a two-step process. Ammonia is first converted into nitrite by a group of bacteria called *Nitrosomonas*. Subsequently, further conversion of nitrite leads to nitrate by another group of bacteria named *Nitrobacter*. Other genera may also be involved in the nitrification process. For example, *Nitrospira*, *Nitrococcus*, *Nitrosogloea*, and *Nitrosocystis* have been identified as participating in oxidizing ammonia in nitrification (Belser, 1979).

Most nitrifying bacteria are autotrophic and utilize carbon dioxide as the carbon source. For oxidation of ammonia, the biochemical reaction is expressed as:

$$13NH_4^+ + 15CO_2 \rightarrow 10NO_2^- + 3C_5H_7O_2N + 23H^+ + 4H_2O \qquad (2.9)$$

where, again, we use $C_5H_7O_2N$ to represent the composition of microorganisms.

For oxidation of nitrite, the reaction expression is written as:

$$NH_4^+ + 5CO_2 + 10NO_2^- + 2H_2O \rightarrow 10NO_3^- + C_5H_7O_2N + H^+ \qquad (2.10)$$

These equations allow the amount of chemicals required for the processes to be calculated.

The growth of nitrifying bacteria is represented by Monod Kinetics:

$$\mu = \frac{\mu_m \cdot S}{k_s + S} \qquad (2.11)$$

where:
  μ is the specific growth rate of the nitrifying bacteria
  μ_m is the maximum specific growth rate
  k_s is the saturation constant
  S is the residual concentration of the growth-limited nutrient.

In the two-step nitrification process, conversion of ammonia into nitrite is limiting reaction; thus it is more convenient to model nitrification on the ammonia-nitrite step, i.e., on the specific growth rate of *Nitrosomonas*, μ_{NS}:

$$\mu_{NS} = \mu_{m \cdot NS} \frac{[NH_4 - N]}{k_s + [NH_4 - N]} \qquad (2.12)$$

Table 2.2 Common values for various Monod kinetic constants applicable to the nitrification process

Monod kinetic constant	Value
μ_{mNS}, 20°C, (pH)$_{optimum}$	0.3–0.5 day^{-1}
k_s	0.5–2.0 mg/L
Y_{NS}	≈0.05 mg VSS/mg [NH$^+_4$–N]
(pH)$_{optimum}$	8.0–8.4

Source: Hultman, 1973. Reproduced with permission of Springer Science + Business Media.

where [NH$_4$–N] is the ammonia concentration, expressed in terms of nitrogen in wastewater in the reactor.

The data for k_s of NH$_4$, the maximum specific growth rate, $\mu_{m \cdot NS}$, and yield in nitrification can be found in Table 2.2 (Hultman, 1973). Once the observed yield constants for NH$_4^-$ and NO$_2^-$ conversions are known, various calculations can be done with Equations (2.9) and (2.10).

2.6 Denitrification

Denitrification can be viewed in some ways as a reversal of nitrification. However, although denitrification does go through a two-step biochemical transformation, its end product is not ammonia or organic nitrogen but, rather, it is gaseous nitrogen (Liu *et al.*, 2003).

Denitrification can only be operated under anoxic conditions when the free oxygen level is very low but not necessary zero, and when a carbon source such as methanol or settled sewer (which has low dissolved oxygen) is available. The biochemical reactions characterizing the denitrification process are brought about by a wide range of bacterial genera – mostly facultative anaerobes such as *Pseudomonas* (*P. fluorescens, P. aeruginosa, P. denitrificans*) and *Alcaligenes*, with *Achromobacterium, Denitrobacillus, Spirillum, Micrococcus,* and *Xanthomonas* often present in wastewater streams (Painter, 1970; Tiedje, 1988).

The overall stoichiometric equation for denitrification using methanol as the carbon source is, according to McCarty *et al.* (1969):

$$NO_3^- + 1.08CH_3OH + H^+ \rightarrow 0.47N_2 + 0.065C_5H_7O_2N$$
$$+ 0.76CO_2 + 2.44H_2O \qquad (2.13)$$

The maximum specific growth rate of nitrifying bacteria or denitrifiers (μ_m) is affected by nitrate and methanol concentrations, temperature, and

Table 2.3 Common values for various Monod kinetic constants applicable to the denitrification process

Monod kinetic constant	Value
$\mu_{m, 20°C, \text{organic matter}}$	3–6 day^{-1}
$\mu_{m, 20°C, \text{methanol}}$	5–10 day^{-1}
$k_{s, \text{MeOH}}$	5–10 COD mg/L
$k_{s, \text{COD}}$	10–20 COD mg/L
$k_{s, \text{NO3}}$	0.1–0.5 O$_2$ mg/L
Y_{MeOH}	0.5–0.65 mg COD/mg COD
Y_{COD}	1.6–1.8 mg COD/mg NO$_3^-$–N

Source: Henze et al., 2001. Reproduced with permission of Springer Science + Business Media.

pH. The growth rate of denitrifiers, μ_D, is represented as a double Monod expression:

$$\mu_D = \mu_m \cdot \left(\frac{D}{k_D + D}\right)\left(\frac{M}{k_m + D}\right) \tag{2.14}$$

where:
  D is the nitrate concentration (mg/L)
  k_D is the half-saturation constant for nitrate (mg/L)
  k_m is the half-saturation constant for methanol (mg/L)
  M is the methanol concentration (mg/L).

Denitrification rate is related to the growth rate of denitrifiers as:

$$q_D = \frac{\mu_D}{Y_D} \tag{2.15}$$

where:
  q_D is the nitrate removal rate (mg NO$_3$ – N mg VSS^{-1}d^{-1})
  Y_D is the yield (mg VSS per mg NO$_3$ – N removed).

Table 2.3 lists some values of the constants in Equation (2.14).

2.7 Further reading

Barnes, D. & Bliss, P.J. (1983). *Biological Control of Nitrogen in Wastewater Treatment*. Spon, London.

Grady, C.P.L. & Lim, H.C. (1980). *Biological Wastewater Treatment: Theory and Applications*. Marcel Dekker, New York, NY.

2.8 References

Belser, L.W. (1979). Population ecology of nitrifying bacteria. *Annual Reviews of Microbiology* **33**, 309–333.

Hultman, B. (1973). Biological nitrogen reduction studies as a general microbiological engineering process. In: Linder, G. & Nuberg, K. (eds.) *Environmental Engineering*. D. Reidel, The Netherlands.

Liu, S.X., Hermanowicz, S.W. & Peng, M. (2003). Nitrate removal from drinking water through the use of encapsulated microorganisms in alginate beads. *Environmental Technology* **22**, 1129–1134.

McCarty, P.L., Beck, L. & P. St. Amant. (1969). Biological denitrification of wastewaters by addition of organic materials. In *Proceedings of the 24th Industrial Waste Conference*, PP1271–1285. Purdue University.

Painter, H.A. (1970). A review of the literature on inorganic nitrogen metabolism. *Water Research* **4**, 393–450.

Tiedje, J.M. (1988). Ecology of denitrification and dissimilatory nitrate reduction to ammonium. In: Zehnder, A.J.B. (ed.). *Environmental Microbiology of Anaerobes*, pp. 179–244. John Wiley and Sons, New York, NY.

3
Physicochemical wastewater treatment processes

3.1 Introduction

Physicochemical processes of wastewater treatment are most evident in the primary treatment facilities of a wastewater treatment plant. The adjective "primary" may have had its undisputed claim on its preeminence in the majority of wastewater treatment plants in the past, but this is no longer the case as the environmental regulations and discharge standards of wastewater effluents are tightening in many developed nations. Nevertheless, primary treatment processes in many wastewater treatment facilities in the US or elsewhere are the commonest wastewater treatment processes, and they contain many forms of physicochemical processing that may be categorized into a small number of unit operations.

Physicochemical processes in the primary treatment of food and agricultural wastewater are intended to remove particulates and other coarse materials from the wastewater stream prior to the secondary treatment processes (mostly biological processes). The removed solids are fed into either aerobic digesters or anaerobic digesters for further volume reduction (Figure 3.1). In primary treatment, only physicochemical processes are used to separate suspended solids and greases from wastewater.

Primary treatment of food and agricultural wastewater usually include screening, flotation, flocculation, sedimentation, and sometimes, granular sand filtration. In a typical wastewater treatment facility for a food processing plant, wastewater is normally held in a tank for several hours, allowing the particles to settle to the bottom and the greases to float to the top. The solids drawn off the bottom and skimmed off the top receive further treatment as sludge, and the clarified wastewater then flows on to the next stage of wastewater treatment. The exact lineup and sequence of unit operations is largely dependent on

---

Food and Agricultural Wastewater Utilization and Treatment, Second Edition. Sean X. Liu.
© 2014 John Wiley & Sons, Ltd. Published 2014 by John Wiley & Sons, Ltd.

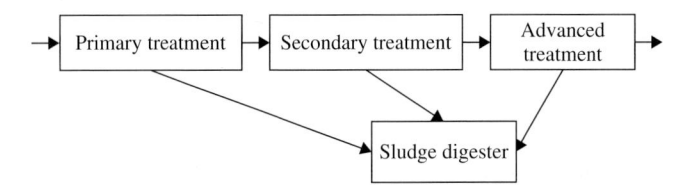

Figure 3.1 A flow chart of a typical wastewater treatment scheme

the characteristics of wastewater streams, objectives of treatment, and local environmental laws and regulations.

3.2 Equalization basins

Generally speaking, flow equalization is not a treatment process or method, but rather a method to improve wastewater treatment processes, whether they are physicochemical processes or biological processes. The purpose of flow equalization is to balance out the process parameters, such as flow rate, organic loading, strength of wastewater streams, pH, and temperature over a 24-hour period. This practice is often applied either at the very beginning of the wastewater treatment plant, aimed at minimizing or controlling fluctuations in wastewater characteristics in order to provide optimum treatment conditions for the subsequent treatment processes in the plant, or at the point right before discharging of effluent. For the former, the target is often focused on the toxicity of the influent, while the latter is aimed at maintaining a predetermined volume of discharge.

Flow equalization usually involves the construction of large basins to collect and hold wastewater streams, from which the wastewater is pumped to treatment facilities at a constant rate. These basins are usually located after pre-treatment facilities such as screens, comminutors, and grit chambers. In the case of industrial wastewater discharging into a municipal wastewater treatment plant, the location of the basins should be placed in the industrial site before discharging, in order to smooth the flow rate and characteristics of the industrial wastewater stream.

Mixing is usually provided to ensure adequate equalization in basins and to prevent the deposition of settleable solids on the bottom of the basins. Additionally, these basins also provide some treatment functions by oxidizing the reduced compounds in the wastewater and reducing BOD through air stripping. The mixing may be achieved by a number of ways, including distribution of inlet flow or baffling, turbine mixing, diffused air aeration, or mechanical aeration.

3.2 EQUALIZATION BASINS

The basic types of flow equalization systems may be categorized as follows:

- *Alternating flow diversion* (see Figure 3.2a): the system alternates filling and discharging one of two flow equalization basins for consecutive time periods. The advantages of this system are constant flow rate and constant pollutant level in the discharging effluent to the treatment facilities. The disadvantage of the system is the high cost of construction, because a large capacity of basins is needed to hold incoming wastewater streams one basin at a time.

- *Intermittent flow diversion* (see Figure 3.2b): the basin is used to divert significant variance in wastewater flow when needed. The diverted stream can be added back to the wastewater stream before it enters the treatment facilities. The diversion tends to run on short time periods.

- *Completely mixed, combined flow* (see Figure 3.2c): this system completely mixes several incoming wastewater streams at the front end of the treatment facilities in a mixing basin in order to level out the variances among the streams. This system works well if the flows are relatively compatible, but any large variance in parameters will create a shock load to the treatment facilities.

- *Completely mixed, fixed flow* (see Figure 3.2d): the difference between this system and the completely mixed, combined flow system is that there is a large holding basin. This not only completely mixes the incoming wastewater streams but also equalizes the flow parameters in the basin before the effluent of the basin goes into the treatment facilities, thus providing a constant wastewater effluent with relatively constant flow parameters.

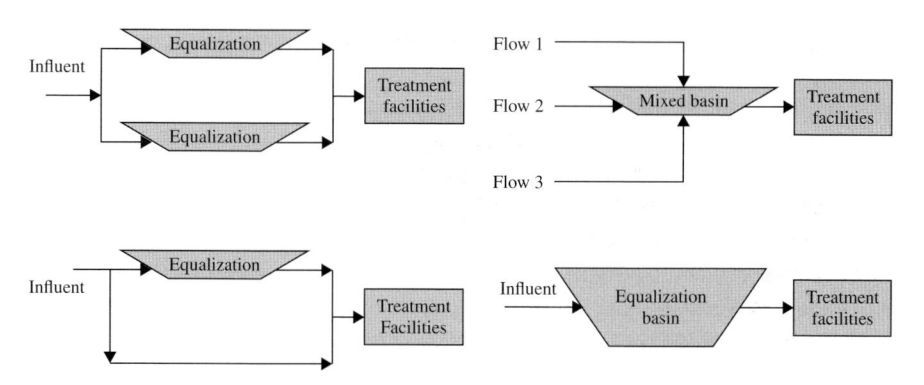

Figure 3.2 (a) Alternating flow diversion equalization system. (b) Intermittent flow diversion system. (c) Completely mixed, combined flow system. (d) Completely mixed, fixed flow system

Design of the equalization facility should begin with an investigation of the characteristics of the wastewater and its variability. A detailed analysis of the pollutants in the stream and flow data collection are essential in order to gain an appreciation of the effect of the nature of the wastewater on downstream wastewater treatment. The most important flow parameters that need to be included in a detailed study of the nature and variability of the wastewater stream are mass flow rate, BOD_5, Total Suspended Solids or TSS, TOC, etc. The data gathered from the study of the wastewater stream tends to a time-series and a statistical analysis is needed to determine the effect of variability on the data of the parameters.

Equalization basins may be designed to achieve flow equalization, concentration of pollutants, or both. For flow equalization, a plot of the cumulative flow volume versus time over 24 hours is compared to the straight line of the average daily flow rate on the same diagram. The equalization volume required is the vertical distance from the point of tangency to the straight line representing the average flow rate, multiplying by a safety factor (e.g., 110%). If the cumulative inflow rate curve goes beyond the line of the average daily flow rate, draw two straight lines that bound the cumulative inflow rate curve and parallel to the average daily rate line. The equalization volume is the distance between the two straight lines that are tangent to the extremities of the curve.

3.3 Screening

Wastewater from food processing or postharvest processing may contain debris, either suspended or floated on the surface. These coarse solids have to be removed at the very beginning of wastewater treatment regimen. Screening of debris is sometimes considered as a "preliminary" treatment, not part of primary treatment of wastewater, but the distinction is more a semantic one than anything else. Screening can be effective in the food industry to reduce the amount of relatively large solids (0.7 mm or larger) in the wastewater quickly and cheaply. The simplest type of screen is an inclined flow-through type of static screen with openings of about 1.5 mm to 6 mm for fine screens and 6 mm or larger for coarse screens, as illustrated in Figure 3.3 (there are also very fine screens with openings of 0.2–1.5 mm which, when placed after coarse or fine screens, can reduce suspended solids to levels near those achieved by primary clarification/sedimentation). In some cases, some sort of scrapping mechanism has to be implemented to avoid clogging of the screens.

Rotary drum screens may also be used in removing coarse solids from food and agricultural wastewater. This type of screen is composed of a drum that rotates along its axis. The effluent enters through an opening at one end and screened wastewater exits from the other end. The retained solids are washed out from

Figure 3.3 Photograph of an inclined screen (see plate section for color version)

the screen into a collector in the upper section of the drum by a spray of the wastewater (Figure 3.4).

The screening media used in screens are generally of stainless steel, with openings ranging from 0.7–1.5 mm. Some of the screening elements may consist of parallel bars or rods (this type of screen is called "bar rack"), wires, grating, wire mesh, or perforated plates. Bar racks are used to protect pumps, values, pipes, and other devices from damages and clogging by large objects in the wastewater stream. Some bar racks are equipped with a mechanical scrapping accessory to clean the screen.

3.4 Flotation

Flotation is a physical process of removing not only oil and grease, but also fine and light suspended particulates from wastewater. Flotation has particular appeal to food wastewater treatment, since this source of wastewater contains substantial amounts of oil/grease floating on the surface. The particulates in wastewater

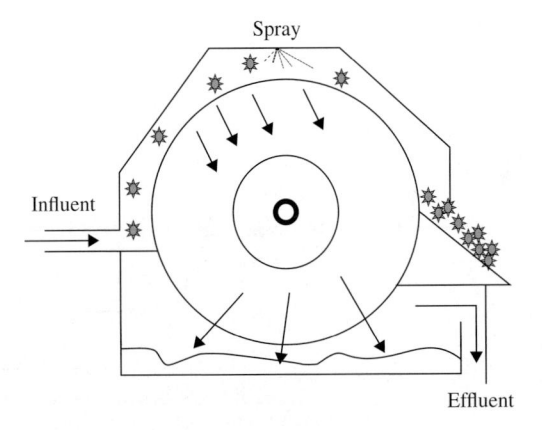

Figure 3.4 A schematic diagram of a vacuum drum screen

that do not settle well and take too much time for settling are also good candidates for flotation treatment.

Flotation is achieved by introducing a gas (usually air) into the wastewater steam, either through pressure-dissolved air in the feed or by direct air diffusers or vacuum. The air bubbles attach themselves to the particulates, causing the particulates and oil/grease to aggregate and rise to the surface, where the particulates are removed by mechanical skimmers. For oil and grease removal, the emulsified oil and grease in wastewater present a problem for the utilization of flotation technology, and it is crucial to break up the emulsion before applying flotation treatment to the wastewater. Adjustments of pH value and/or heat, or the addition of chemicals, are used to destabilize the emulsion.

Dissolved air flotation (DAF) is commonly used in food wastewater treatment. The wastewater feedstock is first pressurized with air in a closed tank and, after flotation, is also used to concentrate sludge from biological processes. Flocculants (see flocculation and coagulation section, below) of inorganic ($FeCl_2$, $FeSO_4$, or $Al_2(SO_4)_3$ and organic (carrageenan, chitosan, and lignosulfonic acid, or their derivatives), as well as polymeric (polyacrylamides) nature are sometimes added to aid flotation operations. As a food wastewater treatment strategy, flotation (DAF, in particular) has its own advantages and disadvantages.

Advantages:

- Reduces oil, grease, and suspended particulates.
- Reduces suspended solids and settling solids if equipped with a bottom sweep.
- Does not require excessive maintenance or management.

3.4 FLOTATION

Figure 3.5 A dissolved air flotation system (see plate section for color version)

Disadvantages:

- Does not remove the BODs associated with soluble materials.
- Disposal or/and treatment of floats is necessary.
- Capital and operating costs could be high.

In a DAF system, wastewater is first mixed with flocculants and pressurized to a pressure of several atmospheres, followed by the release of pressure to atmospheric level by a valve. As the minute bubbles (in the order of 50–100 microns) resulting from de-pressurization rise up to the surface, the particles, oil or grease adhere to the air bubbles to be carried upward. A mechanical skimmer on the top then removes the float. A bottom sweep is sometimes employed to stir up settleable particulates in order to aid flotation operation (see Figure 3.5). For larger flotation systems, a portion (at least 10%) of the effluent from the DAF system is recycled and mixed with the fresh feed of the DAF system.

The oil, grease and suspended particulates removed by the DAF system are concentrated in the float and must be used or disposed of properly to avoid pollution. DAF float can be disposed of by:

- Applying it to land (a permit is required); see also Chapter 7 regarding land applications.
- Depositing it in a landfill (this practice is banned in some states); regulations usually require that the material be easily handled by the machinery in the landfill.

- Using as ingredients for animal feeds; however, the flocculants used must be approved by the FDA, and the float often contains too much liquid for most direct feeding applications or for uses as ingredients for animal feed manufacturing (high energy costs for drying).
- Mixing with sludge for further treatment.

3.5 Sedimentation

Sedimentation is the most common physical unit operation in wastewater treatment, and more so in primary treatment, where sedimentation is the workhorse of the process. The term "sedimentation" is also called "settling" in some of the literature. Sedimentation is, in a nutshell, a process by which the suspended solids, which have higher densities than that of water, are removed from wastewater by the action of gravity in the bottom of the settling tank or basin (also called a clarifier) within a reasonable period of time. Sedimentation basins are usually rectangular or circular, with a radial or upward water flow pattern. Sedimentation is not limited to primary treatment; there is also secondary sedimentation, by which settleable solids in the biological secondary treatment processes are removed. For example, recovery of activated sludge for recycling is achieved with secondary sedimentation.

In a typical wastewater treatment plant, the wastewater stream exiting from screening devices (and after flotation basins) enters the second section of the primary treatment of wastewater treatment or the sedimentation tanks/basins. Here, the sludge (the organic portion of the sewage) settles out of the wastewater and is pumped out of the tanks. Some of the water is removed in a step called thickening, and then the sludge is processed in large tanks called digesters.

Sedimentation uses gravitational force to separate unstable and destabilized suspended solids from wastewater. It is based on the density difference between the bulk of the liquid and the solids. Stabilized (suspended) fine solids, such as colloids, can be destabilized with flocculants (see the section below for flocculation). Sedimentation is also used in biological treatment, such as activated sludge and trickling filters for solid removal. The settling characteristics of the solids are determined by the types of the settling solids and their concentrations.

Sedimentation has four distinct types of settling:

- Discrete settling (Type I), which is the settling of a dilute suspension of solids that do not aggregate.

- Flocculent settling (Type II), which is settling of the particulates that aggregate among themselves and/or with added flocculants to form larger particulates, therefore resulting in faster settling. The sedimentation operation in a typical primary treatment of wastewater operates in this mode.

- Zone settling (also called hindered settling, Type III), which occurs when particulates adhere together, forming a mass that settles as a blanket with a distinguishable interface with the liquid above it.

- Compression zone (Type IV), which occurs when sinking particulates accumulate at the bottom of the sedimentation tank/basin, forming a compressed structure that supports the weight of the particulates that settled in the bottom of the tank/basin.

Although sedimentation basins in primary treatment are characterized by Type II flocculant settling, each of these zones has different characteristics that warrant further analysis.

3.5.1 Discrete settling (Type I)

The settling of non-aggregated solids in a dilute suspension can be described by its settling velocity of individual particulates. In a settling tank/basin, the settling of a discrete particle is not affected by the other particles and is only a function of the fluid property and the characteristics of the particle. This is further illustrated in Figure 3.6, when the movement of the particle of interest is subject to combined effect of the gravitational force downward and the bulk flow toward the outlet:

$$v_t = \text{(tank depth)}/\text{(residence time)} \qquad (3.1)$$

or mathematically

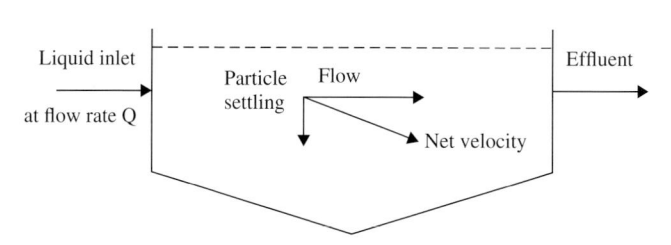

$$v_t = H/t \qquad (3.2)$$

Figure 3.6 Schematic diagram of discrete settling

where:
> H is the depth of the sedimentation tank
> T is the residence time of the particle in the liquid in the tank.

If we assume that the residence time of the liquid is the residence time of the liquid in the tank, then t is given by:

$$t = HA/Q \tag{3.3}$$

where:
> A is the cross-sectional area of the tank
> Q is the overall volumetric flow rate through the tank.

Here, Q/A is the overflow rate of the liquid passing through the tank. So:

$$v_t = Q/A \tag{3.4}$$

The terminal velocity of the particle is equal to the overflow rate of the tank.

The above derivation may strike some readers who have some exposure to fluid mechanics as being suspicious. Since the terminal velocity of a discrete particle in the liquid follows Stake's law:

$$v_t = [g(\rho_s - \rho_l)d^2]/18\mu \tag{3.5}$$

Where:
> d is the size (equivalent diameter of the particle)
> g is the gravitational acceleration
> μ is the viscosity of the liquid
> $(\rho_s - \rho_l)$ is the density difference between the particle and the liquid.

This would suggest that the terminal velocity is function of size and the density of the particle, which is nowhere to be inferred from Equation (3.2) – Equation (3.4). Note that we have assumed that the residence time of the particle in the tank is the same as that of the bulk liquid, or the equivalent depth for the particle settling in the tank is the same as the depth of the tank. The overflow rate of the tank as the terminal velocity of a particle in the tank represents the critical velocity of an ideal particle in an ideal settling tank, assuming:

- the number of ideal discrete particles and the velocity vectors of the liquid are uniformly distributed,
- the liquid flows in the tank as an ideal slug, and
- any particle reaches the bottom of the tank is effectively removed (no re-suspension) (Canale & Borchardt, 1972)

Any ideal particles having terminal velocity v (average velocity among all particles present) greater than v_t is 100% removed from the settling tank/basin. For those particles with less than v_t average terminal velocity, the portion of the particles removed in the tank is equal to v/v_t.

In reality, the discrete settling is more likely associated with hard particulates with high density and size, such as grit and sand – a rare particulate type in a typical food wastewater stream, but which may occur in some sources of agricultural wastewater that is subject to intrusion of soil and dirt.

3.5.2 Flocculent settling (Type II)

Flocculent settling is used in primary clarifiers and the upper zones of secondary clarifiers. Here, the particles in the relatively dilute suspension coalesce or flocculate to form larger particles or aggregates during settling, thus increasing the mass of settling solids as well as the settling velocity (and removal rate). In many food wastewater treatment situations, except very dilute ones, suspended solids cannot be described as discrete particles of known specific gravity (a quantity that is the ratio of density of particle to density of water). In general, larger particles settle faster and have a greater tendency to collide with other, slower-settling particles, resulting in the formation of larger particles in a quiescent body of water. However, wind, hydrodynamic shear, and hydraulic disturbance all contribute to further contacts among particles in the tank. Furthermore, the greater the depth of the tank, the higher the frequency of collisions among particles will be during settling. Therefore, the flocculent settling is dependent on the properties of particles and the liquid, as well as depth of the settling tank/basin.

The settled solids in the bottom of the tank are usually removed promptly, so the greater rate of settling as a result of aggregation of individual particles translates into a greater rate of solid removal from the wastewater. Evaluation of a wastewater stream slated for sedimentation tank or basin is carried out using a settling column, as depicted in Figure 3.7. The laboratory of the settling column is about 15 cm (6 in) in diameter and 305 cm (10 ft) tall, and has several sampling ports that are 61 cm (2 ft) apart. The settling evaluation is conducted by first placing a known quantity of wastewater sample in the column.

Uniformity of both particle size from top to bottom of the column in the beginning of the evaluation, and of the temperature of the liquid throughout the evaluation, should both be accomplished. The wastewater containing suspended solids is allowed to settle under quiescent conditions, small samples of suspension at different ports with pre-set depths are drawn, and concentrations of particles in the samples are determined over pre-set time intervals. The fraction removal of the particles is calculated for each sample analyzed and is plotted against time

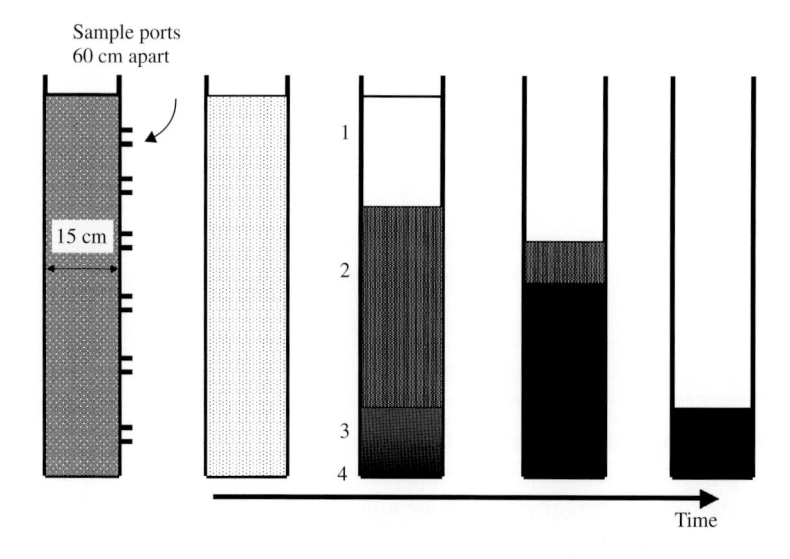

Figure 3.7 Diagram of settling column and zoning settling process

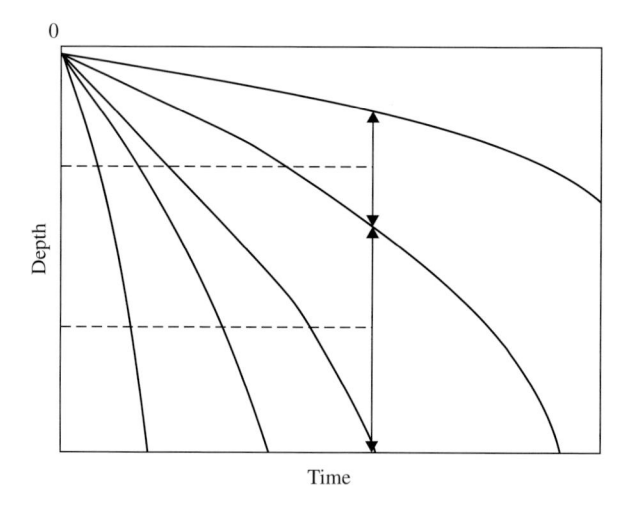

Figure 3.8 Fraction of removal of flocculating particles at each depth

and depth. The fraction of the particles removed at each depth is constructed as a curve line called an iso-concentration line, as illustrated in Figure 3.8. These lines represent the most efficient particle removal loci for a given removal rate. The ratio of the depth to time is the average settling (terminal) velocity of the particles under a given percentage removal.

3.5.3 Zone settling (Type III)

Zone settling, also called hindered settling, acquires its name from the fact that aggregated particulates of a concentrated suspension (activated sludge or flocculated colloids) in the sedimentation basin tend to form a massive, blanket-like suspension with a distinct interface. Zone settling is mainly used in secondary clarifiers. Many wastewater treatment process designers use a batch settling test to determine the interface.

3.5.4 Compression zone (Type IV)

Compression settling involves the highest concentration of suspended solids and occurs in the lower reaches of clarifiers. The particles settle by compressing the mass of the particles beneath them. Compression occurs not only in the lower zones of secondary clarifiers, but also in sludge thickening tanks.

3.6 Coagulation and flocculation

Many substances in wastewater vary greatly in size, from a few Angstroms for soluble solids to a few hundred microns of suspended materials. Consider a force balance upon a clay particle with diameter of 1 micron; in the absence of electrostatic forces, the terminal settling velocity of this particle in water is approximately 10^{-4} cm/s, based on the expression:

$$v = \frac{d^2 g(\rho_s - \rho_f)}{18\mu} \qquad (3.6)$$

where:
  $(\rho_s - \rho_f)$ is the density difference between the particle and fluid (water)
  μ is the viscosity of fluid (water).

This is obviously too low for any practical sedimentation process. To remove a large portion of these substances from wastewater by sedimentation or filtration, smaller particulates need to be aggregated into larger and more readily settleable or filterable particles. This process of forming aggregates from smaller particulates is called coagulation. Many unsettleable small particulates are colloids that can exist stably in water under favorable conditions. The objective of coagulation is to destabilize the colloidal dispersion in wastewater by using either chemical/polymer agents or hydrodynamic forces. The aggregation of colloidal particulates can be visualized as a two-step process: particle contacts/collisions brought up by hydrodynamic forces, followed by particle destabilization to allow particles to attach to one another when they collide.

There are two terms describing aggregations of colloidal particles: coagulation and flocculation. Depending on the industry that employs the unit operation, these terms might mean different things, or they can be synonymous. In environmental engineering, and particularly the wastewater treatment field, "flocculation" refers specifically to the destabilization of colloidal particles by forming aggregates of colloids with added water-soluble polymers (polymer bridges). "Coagulation", on the other hand, is caused by destabilization of colloids through the compression of electrical double layers of the particles. However, this view is not universally shared. Some experts refer to the initial step of adding chemicals (coagulants or flocculants) to the wastewater in an intensely stirred tank as "coagulation" and subsequent slow stirring of the destabilized colloidal suspension in another tank to promote floc growth as "flocculation".

The conventional practice of initial rapid mixing, followed by slow stirring of coagulation-flocculation, is intended to maximize the extent of the formation of aggregates of colloids, but this does not always have desirable outcomes. The slow stirring of the later stage of flocculation produces many large, fluffy flocs with high interstitial water. Although these large flocs are settleable, this could put a severe strain on the operation of dewatering the sludge and may ultimately result in large amounts of sludge destined for landfill (Liu, 1995). Glasgow & Liu (1995) proposed a scheme of coagulation-flocculation that involves slow stirring, interspersed with short bursts of highly intensive mixing in a flocculator, in order to produce flocs with more compact structures.

Many food wastewaters contain large amounts of organic materials, such as proteins, that are colloidal in nature. These tend to be charged as a result of ionization of carboxyl and amino groups or their constituent amino acids and, therefore, stabilizing in the streams. Other organic substances, also common in some wastewaters, may contain grease and oil and become charged due to adsorption of anions such as hydroxyl ions. Destabilization of the colloidal suspension containing these charged colloids requires overcoming the zeta (ζ) potential of the colloid dispersion in order to form aggregates. Zeta potential refers to the electrostatic potential, generated by the accumulation of ions at the surface of the colloidal particle, that is organized into an electrical double layer consisting of the immovable Stern layer and the diffuse layer. The usefulness of the zeta potential as a process parameter is questionable in real-world situations, because it varies with the composition of the suspension and is hardly ever repeatable.

Destabilization of colloids can be achieved through the addition of chemical agents, including charged or nonionic water-soluble polymers. Depending on the

3.6 COAGULATION AND FLOCCULATION

conditions under which the agents are used and the characteristics of the agents, destabilization of colloids in water may be achieved through one or more of four distinct methods:

1. Compression of the diffuse layer of the electric double layer.
2. Adsorption of agents to produce charge neutralization.
3. Enmeshment of colloids in a precipitate.
4. Adsorption of polymeric agents to allow interparticle bridging.

The electric double layer is the result of dynamic charge equilibrium between the particle and water with a zero net electrical charge in the colloidal dispersion (particles plus water). The double layer consists of the charged particles and counterions in water that are attracted to the charges on the particles. The concentrations of counterions near the particles are determined not only by the charges on the particles, but also by the diffusional force due to the concentration gradient in water. The result is a concentration distribution of counterions near the charged particle, with the highest concentration of counterions near the particle surface, decreasing gradually with increasing distance from the particle surface. When the concentration of counterions is low, the electric double layer is extended (because a large volume of the diffuse layer is needed in order to maintain the electrical neutrality of colloidal dispersion).

Destabilization of colloidal dispersion, by adding counterions in the form of chemical agents, is achieved by reducing the volume of the diffuse layer needed in order to maintain electrical neutrality of the colloidal dispersion. The compression of the double layer enhances the likelihood of aggregation among colliding particles, because the attractive forces (van der Waals forces) between particles are short-distance forces that operate only during collisions or near misses. Adsorption of counterions on particles to neutralize the charges on the surface eliminates electrostatic repulsions among colloids, making aggregation possible.

Enmeshment of colloids in wastewater is mainly attributed to precipitation of insoluble $Fe(OH)_3$ or $Al(OH)_3$ when the common coagulants $FeCl_3$ or $AlCl_3$ are added into wastewater under alkaline condition.

When water-soluble polymers are used for destabilizing colloidal dispersion, the mechanism of destabilization is not charge neutralization; since the most effective polymeric coagulants are anionic polymers, even the majority of colloids are negatively charged. The bridging theory stipulates that the chemical groups of the polymer chains bond to the sites of colloidal particulates, forming particle-polymer-particle aggregates (hence the term "bridging" in reference to the mechanism).

Many common polymers used for wastewater treatment are classified as polyelectrolytes, because they contain ionizable groups such as carboxyl, amino, and sulfonic groups. Polyelectrolytes can be positive, negative, or ampholytic (with both positive and negative groups), and they can be prepared from acidic and basic vinyl monomers, from sulfobetaine monomers, from ion-pair co-monomers, or from charged anionic and cationic monomers mixed in varying proportions.

Flocculation is not limited to colloids destabilized by coagulant chemicals and polymers; aggregation of microorganisms is common in biological wastewater treatment plants. It is evident that natural polymers, either excreted by microorganisms or exposed at the surface of the microbial cells, are responsible for enabling this bio-flocculation. These natural polymers are also responsible for destabilizing organic colloids in wastewater. It has been demonstrated that synthetic polymers can also destabilize the microorganism suspension, tending it towards aggregation.

The selection of the optimum type and dosage of coagulant can only be made after judicious experiments with wastewater samples. Many substances can be used as coagulants. For food wastewater containing high proteinaceous substances, it is sometimes required to adjust pH by adding acids or alkalis. For protein-rich wastewater, coagulation of the proteins can be started with denaturing – a process of changing structural conformation of proteins under heat or shear or chemical addition. The downside of denaturing as a way of coagulation of proteins is the high cost associated with its energy requirements; it is cheaper to use chemical agents as coagulants. If the recovered sludge from coagulation-flocculation treatment is to be used for animal feeds, the toxicity of chemical agents as coagulants becomes a very important issue.

Due to the complexity of wastewater compositions and the operational policy, no single set of operating conditions will meet the treatment criteria of food and agricultural wastewater. It is therefore necessary to evaluate coagulants, pH, coagulant dosage, and operational procedures using a laboratory test that simulates the operation of a full-scale coagulation-flocculation, called a jar test. This is scripted lab testing, conducted using a series of beakers and stirrer in a jar test apparatus (Figure 3.9).

Jar tests have been used to evaluate the effectiveness of various coagulants and flocculants under a variety of operating conditions for water treatment. The procedures and evaluation process have been adapted to dredged material. However, conducting jar tests and interpreting the results to determine design parameters are not simple tasks, because there are many variables that can affect these tests. Only experience can assist in applying the following jar test procedures to a specific project.

3.6 COAGULATION AND FLOCCULATION

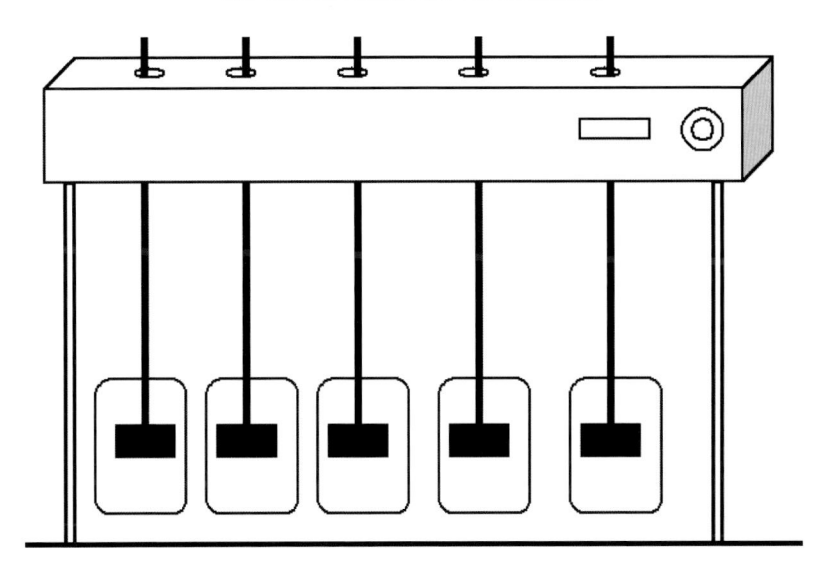

Figure 3.9 A schematic diagram of jar test apparatus

Jar tests are used in these procedures to provide information on: the most effective coagulant; optimal coagulant dosage and feed concentration; the effects of dosage on removal efficiencies; the effects of concentration of the suspension on removal efficiencies on removal efficiencies; the effects of mixing conditions on removal efficiencies; and the effects of settling time on removal efficiencies.

- Fill the jar testing apparatus containers with sample wastewater from a stock suspension (either a real sample or synthetic one with composition similar to the wastewater) of known turbidity, color, alkalinity, and pH. Calculate the amount of alkalinity required to react with the maximum dosage of aluminum or ferric sulfate. If necessary, augment the natural alkalinity by the addition of 0.1 N Na_2CO_3, so that the alkalinity will be at least 0.5 meq/L (25 mg/L as $CaCO_3$). One container will be used as a control, while the other containers can be adjusted depending on what conditions are being tested. For example, dosage of coagulants, pH, and settling time in the containers can be adjusted to determine the optimal conditions.

- If it is a test for an existing coagulation-flocculation process, the procedure should reflect the actual conditions of the specific plant in terms of rapid mixing RPM and time, slow mixing RPM and time, and, finally, settling time. If not, choose a set of appropriate stirring speeds for rapid and slow mixing, mixing times and settling times for flocs to settle completely.

Add chemicals (aluminum or ferric sulfate) to each beaker near the vortex at high RPM for a minute and follow the actual or proposed operating conditions. Next, look at the beakers and determine which one has the best results (if any). An under-dosed suspension will cause the sample to look cloudy, with little or no floc. An overfeed suspension will cause fluffy flocs to occur and will not settle well. The beaker with an appropriate dosage of coagulant will have floc that has settled to the bottom and the water above, and it will be clear, determined either by vision or a nephelometer. If none of the beakers appear to have good results, then the procedure should be re-run using different dosages until the correct dosage is determined.

- The above jar test procedure follows a conventional empirical experiment design, where variables are explored one factor at a time, keeping the other factors constant. This is not always effective or practical, because the optimal conditions identified by the conventional approach may not be the true optimum if interactions between factors are present. Variables such as pH, mixing, and stirring speed may be important factors and should be included in the experiment design; factorial design method may yield a better test result.

3.7 Filtration processes

Filtration is often employed in wastewater treatment, with or without prior treatments by coagulation-flocculation and sedimentation. It is used for removal of flocs (or bioflocs) from primary and secondary wastewater treatment processes; for solids remaining in effluents from primary and secondary wastewater treatment processes; and for precipitates from physicochemical treatment of phosphate from an advanced wastewater treatment stage. Earlier applications of filtration for wastewater treatment borrowed heavily from design and operational experience with potable water treatment, but the wastewater treatment methods have been perfected since that time. Filtration of sludge on rotary vacuum filters for dewatering is a common application of filtration in wastewater. Other applications of filtration include dewatering of digested sludge on sand beds and wastewater treatments with deep granular filters (sand, dual-medium, and multimedia) after most of solids are removed.

Wastewater treatment using filtration is usually designed to get rid of microorganisms, to reduce turbidity and color, to remove odors, to reduce the amount of iron, and to remove most other solid particles that have remained in the water. Water is sometimes filtered through activated carbon particles (see also the section on Adsorption below), which removes refractory organic particles.

Filtration in wastewater treatment is a unit operation that mirrors the natural system that treats impaired waters. Groundwater is filtered through layers of sand or/and soil (underground strata) and potable quality of water may be obtained in wells deep beneath the surface. Naturally, the filtration media used in wastewater treatment are sand, crushed anthracite coal, diatomaceous earth, perlite, and powdered or granular activated carbon. Combination of different media are commonly used; a dual-media system may comprise coal over sand, while a multimedia filter may be layered with garnet sand, silica sand, and coal.

Filtration involves complex mechanisms, because these depend on the physicochemical characteristics of the suspended materials and the filter medium, the rate of filtration, and the composition of the wastewater. Deep granular filters consisting of 18–30 inches (46–76 cm) of filter medium supported on an underdrainage system; the filter may or may not be open to the atmosphere. A closed filter usually involves pressure and is called a pressure filter, while open ones are termed as gravity filters. Precoat filters comprise a number of porous septa in a filter housing connected to a collection manifold. The septa hold a thin layer of filter medium that is deposited hydraulically on the outside of the septa at the beginning of the filtering cycle. The filter septa are either open to the atmosphere while it is submerged in a tank or totally enclosed in a pressure tank. These are sometimes called vacuum filters.

The characteristic designs of different types of filters contribute to the mechanisms of filtration in the systems. For example, precoat removes impurities at the surface through a formation of filter cake made of the impurities, and therefore is a mechanism of straining water. In deep granular filters with coarse materials, on the other hand, removal of particulate materials from wastewater is primarily within the filter bed, as wastewater moves through the spaces formed by the filter medium. This type of filtration, as seen in some deep granular filters, is called depth filtration. However, the actual removal of impurities involves several mechanisms, including interstitial straining, gravitational settling, diffusion, interception, hydrodynamic interactions, and attachment/adsorption due to electrostatic interactions, chemical bonding, or specific adsorption, which may be affected by the type and dosage of the coagulant used prior to filtration. The removal results in many deep granular filters may be dominated by a combination of surface cake removal and depth removal mechanisms.

The flow of wastewater through a filter at the rates commonly used in wastewater treatment is hydraulically the same as flow through underground strata in a natural system of groundwater treatment. The rate of flow through a filter may be expressed as follows:

$$\text{Rate of Flow} = \frac{\text{Driving Force}}{\text{Filter Resistance}} \tag{3.7}$$

where the driving force represents the pressure drop (head loss) across the filter.

Historically, single-phase wastewater flow through porous media has been modeled using either the linear Darcy's law, or some empirical nonlinear relationship between the pressure gradient and the Darcy velocity as an approximation to momentum conservation. For example, head loss through granular materials in the laminar range may be described by Darcy's law:

$$v = K_p \frac{h_f}{L} \tag{3.8}$$

where:
  K_p is coefficient of permeability
  h_f is head loss
  L is depth (or length) of filter.

The coefficient of permeability can be determined experimentally. Theoretical relationships have been developed to calculate coefficient of permeability from measured physical characteristics. Non-linear relationships between the head loss and the Darcy velocity v have been used to illustrate important variables attributing to the head loss, such as the following:

$$\frac{h_f}{L} = J \frac{\nu(1-\varepsilon)^2}{g\varepsilon^3} v \left(\frac{\sigma_s}{d_p}\right)^2 \tag{3.9}$$

where:
  J is a constant of approximately 6 in a laminar flow
  ε is porosity ratio of the filter bed
  d_p is diameter of the particles
  σ_s is the shape factor
  ν is fluid kinematic viscosity.

Values of shape factor range from 6.0 for spherical particles to 8.5 for granular materials such as anthracite coal.

Filter effluent quality patterns vary with the characteristics of the filter medium, the solids, the water chemistry, and the operating conditions of the filter. For precoat filters, the effluent quality is usually excellent, particularly after formation of the filter cake, which acts as an additional filtering medium.

Removal of impurities by granular filter usually occurs within the cavities and interstices of the filter medium. Formation of cakes in these filters may contribute to the filtering of solids. Since the filtering occurs within the granular bed, the burden of removal shifts gradually from the upper layers to the lower layers; it is observed that a granular filter usually starts with high removal efficiency, but degrades steadily over time as layers of the filter medium become progressively saturated with pollutants.

Filtration in wastewater treatment is a unit operation that frustrates designers of wastewater engineering systems because of the inherited variability in the characteristics of the feed stream to be filtered. This is not only a testament to the diversity of wastewater sources, but also to the fact that the filtration process is often employed as a supplemental process to the primary or secondary wastewater treatment processes. It is thus subject to changes in the degree of coagulation-flocculation of colloidal dispersion in the settling basins or tanks, as these changes lead to variability in particle size and distribution. It is often necessary to conduct pilot plant studies to ensure that the filter configuration selected for a given task will perform adequately for the treatment objective.

3.8 Adsorption

Adsorption is a physicochemical process that generally occurs at the interface of fluid–solid phases and is sometimes used to remove certain species that cannot otherwise be effectively removed from the wastewater stream. Although it is quite possible that liquid-liquid or gas–liquid interaction results in adsorption, it is more common in wastewater treatment to observe interactions of two fluid phases as absorption. Adsorption and absorption belong to the same category of physicochemical processes called *sorption*, but they are two different phenomena. The main difference between them lies in the equilibrium phase they eventually form: if it is homogeneous, it is absorption, otherwise it is adsorption. Adsorption can be either physical (such as binding caused by van der Waals forces) or chemical/activated (such as covalent bonds). It is quite possible that both styles of sorption exist in wastewater treatment employing adsorption, since a variety of substances could be in the stream.

In general, the materials being adsorbed are called adsorb*ates*, while the materials adsorbing adsorbates are termed adsorb*ents*. Majority of the adsorbents in wastewater treatment are solids. These solids can be roughly divided into three categories: carbons; inorganic materials; and synthetic polymers. Carbon-based adsorbents, such as activated carbon, are the commonest adsorbents in use in wastewater treatment plants. They are effective and relatively inexpensive, and they even can be made from agricultural wastes. For example, peanut shells from the Deep South of the USA are currently being tested for use as adsorbents for heavy metal and other toxic material removals. Other sources of raw materials for carbon-based adsorbent manufacturing include petroleum coke, wood, and coconut shells.

Inorganic materials used for adsorption include classes of activated alumina and zeolites. These materials are expensive and are more size- and/or species-specific. They are more likely to be found in fine chemical manufacturing, pharmaceuticals, and other high-profit margin industries.

Synthetic polymers can be used as adsorbents, as they are easily functionalized on the surfaces. Ion exchange polymers target ions in the solution, so they are good candidates for removing ions in water purification or in wastewater treatment to remove certain heavy metal or other toxic ions. Polymers that do not rely on ion exchange mechanism utilize hydrophobic or hydrophilic interactions (a stronger form of molecular force than van der Waals forces) to adsorb the adsorbates.

In order to describe the capability of an adsorbent adsorbing certain adsorbate species, we often examine an isotherm of the adsorbent – an adsorption data plot of the amount adsorbed per unit mass of the adsorbent versus concentration of adsorbates in the solution. There are four common isotherms (curves) typical of all adsorptions: linear, Freundlich, Langmuir, and "unfavorable", as shown in Figure 3.10.

The last named of these has an upward curvature on the adsorption vs. concentration plot and receives no favor, as its name suggests; therefore, it is so termed

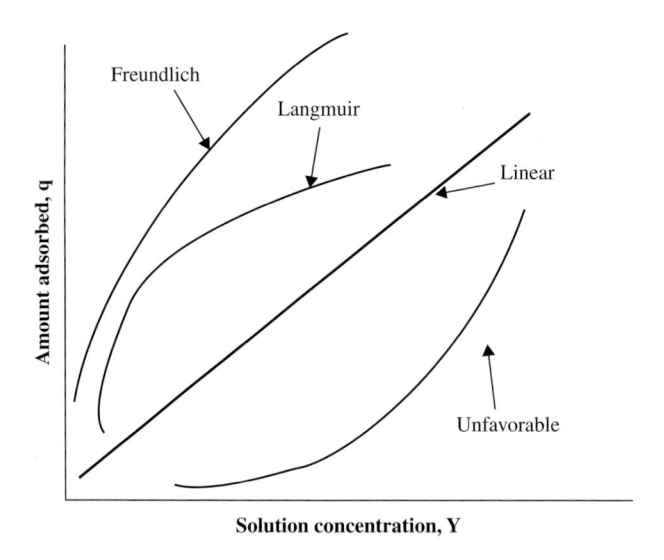

Figure 3.10 Schematic isotherms of adsorption

3.8 ADSORPTION

but never tended. The isotherms reveal not only the capacity of the adsorbent, but also the adsorption patterns and its equilibrium state. The linear isotherm is a simplistic mathematical gimmick, straightforward but unrealistic:

$$q = Ky \qquad (3.10)$$

where K is a proportionality constant associated with equilibrium adsorption.

The reason that the linear isotherm is included in discussions is because it is simple and provides a rather reassuring reference to both Langmuir and Freundlich isotherms.

The Langmuir isotherm is more widespread and easily explained away with theoretical basis that accounts for irreversible adsorption as a result of equilibrium between adsorption and desorption. Its expression is not neat but is insightful:

$$q = \frac{q_0 y}{K + y} \qquad (3.11)$$

where:

q_0 is the total concentration of the adsorption sites on the adsorbents
K is the reciprocal of the equilibrium constant of irreversible adsorption on the sites.

For data processing of the Langmuir type of equations, a linearization step is often all that is needed. For example, re-arrangement of Equation (3.11) below reveals the linear relationship of $1/q$ vs. $1/y$:

$$\frac{1}{q} = \frac{1}{q_0} + \frac{K}{q_0 y} \qquad (3.12)$$

Here, q_0 is considered as a constant.

The Freundlich isotherm is defined as:

$$q = Ky^n \qquad (3.13)$$

where n and K are experimentally determined constants; in many cases, n is less than 1 (but never equal to or larger than 1). Occasionally, for some Freundlich isotherm-type adsorption curves f that have an n value close to 1, the linear isotherm is used for approximation purpose. To use the Freundlich isotherm to fit the experimental data, one has to transform the equation into a linear form by taking logs on both sides of Equation (3.13), resulting in:

$$\log q = \log K + n \log y \qquad (3.14)$$

It is obvious that plot of $\log q$ vs. $\log y$ is linear.

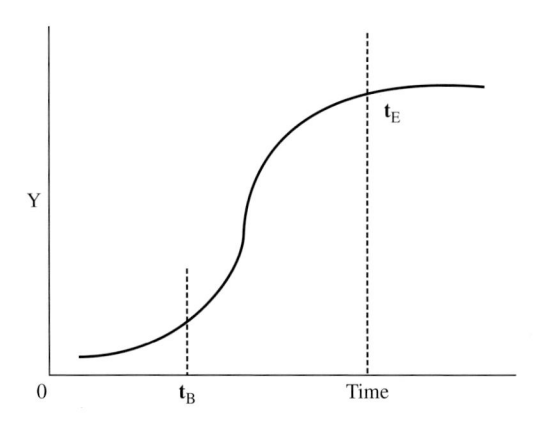

Figure 3.11 Schematic diagram of a breakthrough curve of adsorption

The common way to determine the best fit of the isotherms in an adsorption experimental data is to plot data according to three common isotherms until a best-fit model (isotherm) is found. It is possible that none of the isotherms might fit precisely to the data; in that case, a best fit (least deviated from the majority of data points – a statistical analysis of data fitting of the linearized isotherm) may be needed.

In adsorption operations, adsorbents are often resided in a packed bed. This arrangement, which is akin to a plug flow reactor, has the best mass transfer driving force of adsorption of adsorbates onto the adsorbents in the packed bed. As adsorbates pass through the bed over the time, the adsorbents in the bed extract the adsorbates from the solution and, eventually, the adsorbents in the bed become saturated and the concentrations of the adsorbates from the effluent (called eluent) start to rise, as illustrated in Figure 3.11.

The s-shaped curve of adsorbates' concentrations vs. eluent volume (often expressed as time) that depicts the starting points of sharp rising and leveling-off corresponding to exhaustion of the adsorbents of adsorbates' concentrations leaving the packed bed, is called the breakthrough curve. This is used to compare different adsorbents in a packed bed. In addition to the breakthrough curve obtained from evaluation of adsorbents in a laboratory or provided by the vendor, another issue, called bed efficiency, is also important to practical application of adsorption technology. This is related to the breakthrough curve; the packed bed efficiency over time is a measurement of the effect of the bed (the shell/container holding the adsorbents and the packing factor) on adsorption capacity:

$$\theta = 1 - \frac{t_E - t_B}{2t_B} \tag{3.15}$$

3.8 ADSORPTION

where:

θ is the fraction of the packed bed that is loaded with adsorbates

t_E is eluent time (corresponding to the starting point of leveling-off of the breakthrough curve in Figure 3.11)

t_B is starting point of the breakthrough curve.

This equation, through θ, indicates the actual saturation of the adsorbents in the bed when the breakthrough curve is observed, which is always lower than 100% – hence, the bed efficiency. It should be emphasized that Equation (3.15) is an approximation, because its derivation involves the assumption of the breakthrough curve as a step function – a sharp, vertical rise of the eluent concentration, in contrast to the more rounded curve shown in Figure 3.11. However, it is a relatively adequate approximation, because of the concentration profile developing quickly in the bed and "sharpening" of the curve. Attempting to make a numerical integration over the breakthrough curve does not lead to more accurate results, since the gain from elaborate mathematical treatment of the breakthrough curve will be negated by the errors associated with the numerical methods used, as well as by the curve itself. The approximation method shown in Equation (3.15) is often recommended by adsorbent manufacturers.

Once θ is known, the unused bed length l' can be calculated as follows:

$$l' = l(1 - \theta) \qquad (3.16)$$

where l is the bed length. The used bed length $l - l'$ is therefore saturated as:

$$l_{sat} = \left[\frac{y_0 v}{q_0 (1 - \varepsilon)} \right] (t_E - t_B) \qquad (3.17)$$

where:

v is the superficial velocity of the effluent

ε is the void fraction of the bed

y_0 and q_0 are the initial concentration of the adsorbate and saturated concentration of the adsorbate on the adsorbent, respectively.

For a batch packed bed, if we know the initial and final concentrations of the adsorbate and the amount of the feed that goes through the bed, we can easily calculate the amount of the adsorbent used, based on mass balance on the adsorbate:

$$(y_0 - y)V = qW \qquad (3.18)$$

where:

V is the amount of the feed in volume

W is the weight of the adsorbent.

The adsorption q is in equilibrium with the concentration of the depleted solution, y and can be expressed by one of three isotherms. However, for a packed bed, Equation (3.18) is valid but q is in equilibrium with the feed concentration, y_0, thus enabling much more efficient mass transfer and requiring less adsorbent. This is the reason that packed bed adsorption units, which are close to idealized reactors with plug flow in terms of fluid dynamics and mass transfer, are commonly used.

3.9 Chemical oxidation

Chemical oxidation in wastewater treatment is a process in which undesirable chemical species are converted through oxidation (e.g., redox reactions) to chemical species that are neither harmful nor objectionable. It modifies the structure of pollutants in wastewater through the addition of an oxidizing agent. During chemical oxidation, one or more electrons transfer from the oxidant to the targeted pollutant, causing its destruction. Both organic matter (including microorganisms) and inorganic substances can be subject to oxidation. Examples of organic matter for chemical oxidation include humic acids, phenols, amines and bacteria. Common inorganic substances that are toxic or/and objectionable may include ions such as Fe^{2+}, Mn^{2+}, S^{2-}, CN^-, and SO_3^{2-}. A handful of oxidizing agents are capable of oxidizing the undesirable substances adequately, namely, oxygen, ozone, UV/H_2O_2, Fe^{2+}/H_2O_2 (Fenton's reaction), potassium permanganate, chlorine, chlorine dioxide, and zero-valent iron nanoparticles.

Chemical oxidation applications in wastewater are used throughout entire wastewater management operations, from collection of wastewater streams, through treatment, to disposal. In collection facilities, chemicals such as Cl_2, $FeCl_3$ and O_3 or H_2O_2 are employed in controlling slime growth and corrosion.

In treatment operations on wastewater, chemicals are used in a number of places, often as supplemental procedures for a major wastewater treatment unit operation. For example, Cl_2 and O_3 are applied to reduce BOD and ammonia. Largely, however, chemical oxidation procedures are often used to counter the problems of biological, microbial, or nostrilic (odorous) nature. When the wastewater effluent from a treatment plant is ready for disposal, the effluent is often treated with Cl_2, H_2O_2 or O_3 to ward off bacteria and odor.

In potable water treatment (including membrane-based processes, but with addition of ammonia to neutralize the residual Cl_2 or H_2O_2 in order to protect membranes), chemical oxidation is used for disinfection with Cl_2 (mainly in the US and a number of other nations) or O_3/H_2O_2 (mainly in western Europe). Chlorine has a better residual anti-microbial potency than O_3 or H_2O_2, but it could produce harmful substances as described below.

One common (and inexpensive) method of chemical oxidation, referred to as alkaline chlorination, uses chlorine (usually in the form of sodium hypochlorite) under alkaline conditions to destroy undesirables such as cyanide and some pesticides. However, using alkaline chlorination for chemical oxidation of pollutants may generate toxic chlorinated organic compounds as byproducts, including chloroform, bromodichloromethane, and dibromochloromethane. Adjustments to the design and operating parameters may alleviate this problem, or an additional treatment step (e.g., steam stripping, air stripping, or activated carbon adsorption) may be required to remove these byproducts.

Oxidation by hydrogen peroxide alone is not effective for converting high concentrations of certain refractory contaminants, such as highly chlorinated aromatic compounds and inorganic compounds (e.g., cyanides), because of low rates of reaction at reasonable H_2O_2 concentrations. Transition metal salts (e.g., iron salts), ozone, and UV-light can activate H_2O_2 to form hydroxyl radicals that are strong oxidants:

- ozone and hydrogen peroxide

$$O_3 + H_2O_2 \rightarrow OH\cdot + O_2 + HO_2. \tag{3.19}$$

- iron salts and hydrogen peroxide

$$Fe^{2+} + H_2O_2 \rightarrow Fe^{3+} + OH\cdot + OH-. \tag{3.20}$$

- UV light and hydrogen peroxide

$$H_2O_2[+UV] \rightarrow 2OH. \tag{3.21}$$

The oxidation processes utilizing activation of H_2O_2 by iron salts are referred to as Fenton's reagent.

In general, oxidation processes that are based on the generation of radical intermediates are termed advanced oxidation techniques. Hydroxyl radicals (oxidation potential: 2.8 V) are stronger oxidants than ozone and H_2O_2. Hydroxyl radicals non-specifically oxidize target compounds at high reaction rates of the order of 10^9 M^{-1} s^{-1}.

3.10 Membrane separations

Membrane separations have been playing an increasing role in wastewater treatment. This is most evident in processing whey wastewater with ultrafiltration in the dairy industry – one of many types of membrane separation technologies – and in membrane pervaporation of volatile organic compounds (VOCs) from wastewater. Currently, a new field of membrane separations, called nanomembrane technology, is said to be the future nanotechnology for achieving clean water with this purported type of "smart membranes". It is predicted that nanomembranes are able to separate molecules by differences between molecular weights of a mixture of compounds, a feat that current membrane types (polymeric or inorganic) are unable to achieve. Whether this prediction can really hold water in the future is anyone's guess; however, membrane-based technologies are will undoubtedly be found increasing use among the evolving water and wastewater treatment landscapes. Membrane technology is an important tool to recover valuable materials from food and agricultural wastewaters, as described in a later part of this book.

A membrane can be viewed as a discreet (or discriminating) barrier that allows some components of the wastewater feed to pass through the membrane faster than other components. A membrane provides a third phase – a mostly solid phase that straddles between two fluid phases which serve as the origin and the destination of the separation process. The membrane is most likely polymeric, though new inorganic membranes are now emerging at a rapid pace. The principal mechanisms of membrane separations are molecular diffusion in solids and tortuous viscous flows in microporous solids.

Membrane technology is an evolving separation technology and, because of its multidisciplinary character, it can be used to perform a large number of separations in food and agricultural wastewater treatment. The membrane processes that are commonly found in processing plants or research laboratories include microfiltration (MF), reverse osmosis (RO), ultrafiltration (UF), nanofiltration (NF), electrodialysis (ED), membrane distillation (MD), and pervaporation (PV). Membrane processes are based upon different separation principles or mechanisms, and their applications in food processing range from concentration of food fluids to aromatic flavor recovery.

The membrane itself is at the center of every membrane process. However, membrane separations can only be achieved when a driving force is applied to the underlying membrane process. A schematic diagram of a two-phase conceptual system is shown in Figure 3.12.

It should also be remembered that there no perfect man-made membrane has ever existed. This situation will continue into the foreseeable future until, perhaps, we fully understand the mechanisms that regulate the mass transfer in the membrane, and we are able to tailor the membrane structures to the needs of separating molecules of interest by using the latest advancements in nanotechnology. In assessing membrane systems, two experimental parameters that determine the overall performance of membrane processes should be

3.10 MEMBRANE SEPARATIONS

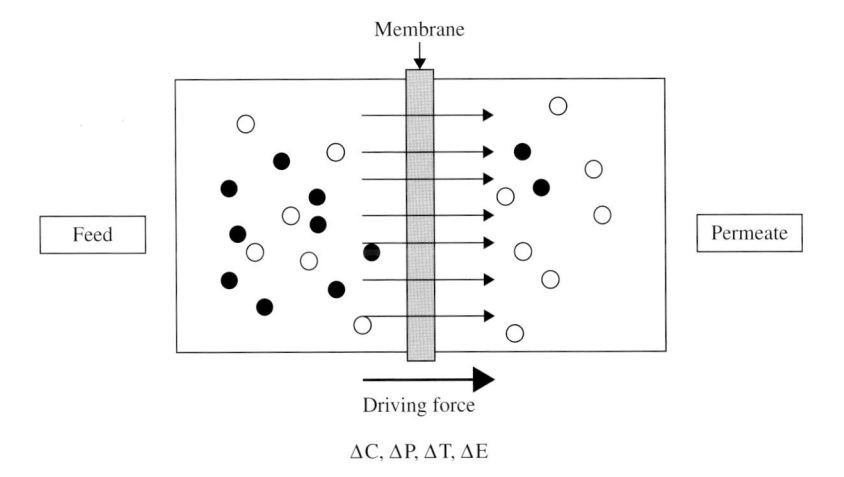

Figure 3.12 Schematic diagram of a conceptual two-phase membrane system

the main focus of designers' attention. The first one is selectivity; the other, permeation flux.

The selectivity of a membrane toward a mixture, which characterizes the extent of separation, is customarily expressed by one of two quantities: the retention, R, and the separation factor, α. The retention, R, is more suitable for the membrane separation of a dilute binary system and given by

$$R = \frac{C_f - C_p}{C_f} \tag{3.22}$$

where:

C_f is the solute concentration in the feed stream
C_p is the solute in the permeate.

The value of R varies between 100% (complete rejection or retention) and 0% (complete permeation). For most mixtures, however, selectivity factor is more adequate:

$$\alpha_{ij} = \frac{(C_i/C_j)^p}{(C_i/C_j)^f} \tag{3.23}$$

where C_i and C_j are the concentrations of components i and j in the permeate and in the feed. The value of α is greater than 1 if the component i is more readily than component j and if the separation occurs.

The other parameter, permeation flux, takes many forms, depending upon the underlying membrane processes. It is normally expressed as:

$$J_i = -K\frac{dg}{dz} \tag{3.24}$$

where:

K is the phenomenological coefficient

dg/dz is the driving force, expressed as the gradient of g (concentration, temperature, pressure) in the z direction toward the membrane.

The phenomenological coefficient, K is strongly related to the driving force, module configuration, and operating conditions.

Membrane processes can be classified according to the nature of their driving forces and the pore size of the membrane. Although all membrane processes are driven by electrochemical potential gradient, one particular driving force usually dominates in a membrane process. Three types of membrane separation processes relevant to the food industry can be considered: those that are driven by hydrostatic pressure difference: those driven by partial vapor pressure gradient; and those driven by electrical potential differences. A brief general description of the membrane processes used or potentially usable in various operations of the food industry are provided in the following sections.

3.10.1 Membrane separation by hydrostatic pressure difference

Membrane performance of a pressure-driven system is usually described by the flow rates of water (solvent) and solute (permeate). The flow of water (volume flux) through a membrane without considering concentration polarization and fouling (or/and gel layer) is expressed by:

$$J_w = K_w(\Delta P - \sigma \Delta \Pi) \tag{3.25}$$

where:

K_w is water permeability

ΔP is the applied pressure

$\Delta \Pi$ is the osmotic pressure difference

σ is the reflection coefficient of the membrane toward the solute, which is a measure of degree of solute rejection.

The driving force is $(\Delta P - \Delta \Pi)$ as dg/dz in Equation (3.24). Since there is no perfect membrane, we may suspect (and this can be verified) that some solutes,

3.10 MEMBRANE SEPARATIONS

including those that are undesirable, also transport across the membrane, though less freely than water:

$$J_s = K_s(C_m - C_p) \tag{3.26}$$

where:
- K_s is the solute permeability
- C_m and C_p are concentrations on the upstream side of the membrane and on the permeate side, respectively.

Note that K_s has a different unit from that of K_w, because the driving force in Equation (3.26) is expressed as the solute concentration difference.

While Equations (3.25) and (3.26) are generally considered as valid phenomenological expressions, the true meanings of C_m and C_p are not what they seem to be. This is because, in a pressure-driven membrane process, the retained solutes transported by convective trans-membrane flux can accumulate at the membrane surface, leading to a high concentration of solutes near the membrane – a phenomenon called concentration polarization. This concentration gradient is encompassed in a region designated as the boundary layer (velocity has its own gradient, due to the viscous effect at the water-membrane interface for a common cross-flow membrane configuration – thus, a velocity boundary layer).

In a steady state situation, the concentration polarization is the result of solute build-up, counterbalanced by the solute flux through the membrane, plus the diffusive flux of solute at the membrane surface toward the bulk flow on the upstream side of the membrane. The magnitude of the concentration polarization is expressed by Equation (3.27), as a result of the solute mass balance based upon the concentration profile of the film model illustrated in Figure 3.13.

$$\frac{C_m - C_p}{C_b - C_p} = \exp\left(\frac{J_s}{k}\right) \tag{3.27}$$

where:
- C_b is the concentration of the solute in the bulk flow
- K is the mass transfer coefficient, that is the ratio of diffusivity of the solute in the solvent to the thickness of the concentration boundary layer, which can be interpreted as the mass transfer coefficient when the permeation flux approaches zero.

The cause of the concentration polarization phenomenon is different in reverse osmosis from that in microfiltration or ultrafiltration. In reverse osmosis, as the low molecular weight material is retained on the membrane surface (see

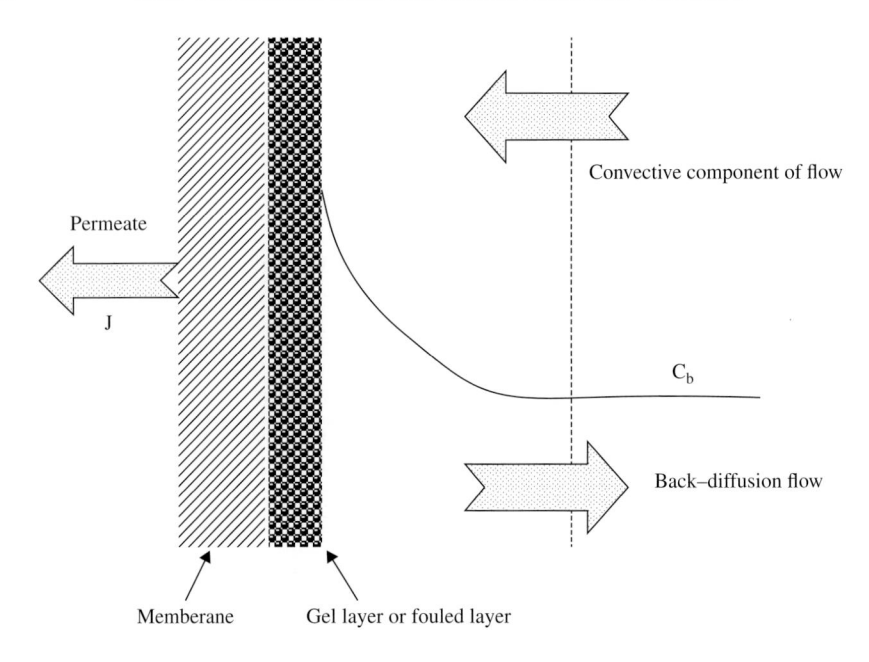

Figure 3.13 A schematic diagram of a film model

the following section about the characteristics of all pressure-driven membrane processes), the increase in the solute concentration causes the osmotic pressure to rise, which decreases the water flux as illustrated in Equation (3.25). In ultrafiltration, the high concentration of larger molecules accumulated on the membrane surface does not result in significant osmotic pressure increase. However, these retained molecules may lead to precipitation and, possibly, the formation of a gel layer on the membrane surface.

The mass transfer coefficient, k, in Equation (3.27) has to be determined experimentally, since the thickness of the concentration boundary layer is usually an unknown quantity that is strongly influenced by hydrodynamics of the system. However, k can often be related to the semi-empirical Sherwood number correlations, with the following form of expression:

$$Sh = \frac{Kd_h}{D} = a\text{Re}^b Sc^c = a\left(\frac{\rho u d_h}{\mu}\right)^b \left(\frac{v}{D}\right)^c \tag{3.28}$$

where:
  Re is the Reynolds number
  Sc the Schmidt number
  Sh is Sherwood number

a, b, and c are all constants.
μ and v are dynamic viscosity and kinematic viscosity, respectively
ρ is the density
D is the diffusivity
d_h is the hydrodynamic diameter.

It is clear in Equation (3.28) that the mass transfer coefficient k is mainly a function of the feed flow velocity, the density, the viscosity, the diffusivity of the solute, and the membrane module type. Many Sherwood relationships for different flow regimes and membrane module shape and dimensions are available in the literature (Cheryan, 1986; Rautenbach & Albreht, 1989; Mulder, 1991).

Microfiltration (MF)

Microfiltration (MF) is a form of filtration that has two common forms. One form is cross-flow separation, where a fluid stream runs parallel to a membrane. There is a pressure differential across the membrane, which causes some of the fluid to pass through the membrane, while the remainder continues across the membrane, cleaning it. The other form of filtration is called dead-end filtration or perpendicular filtration. Here, all of the fluid passes through the membrane, while all of the particles that cannot fit through the pores of the membrane are stopped.

Cross-flow microfiltration is used in a number of applications, as either a pre-filtration step or as a process to separate a fluid from a process stream. Dead-end microfiltration is used commonly in stopping particles in either pre-filtration or final filtration before a fluid is to be used. Cartridge filters are typically composed of microfiltration media. MF is a pressure-driven membrane filtration process that has a membrane with a typical pore size of 0.2–2 µm, which is able to retain particles with molecular weights equal or larger than 200 kDa. MF membranes are symmetric, with a characteristic sponge-like network of interconnecting pores. They have been successfully used in the beer brewing industry to remove bacteria in the production of long shelf-life draft beers. The dairy industry has also found MF useful in removing bacteria or particulate substances and fractionation of milk proteins. MF is the membrane process that most closely resembles to a conventional filtration unit. Its transport mechanism is undoubtedly sieving action; thus, the volume flux through the MF membranes is either expressed with a Hagen-Poiseuille relationship:

$$J = \frac{\varepsilon r^2 \Delta P}{8 \mu \tau \Delta z} \qquad (3.29)$$

if the membrane is perceived as a bunch of straight capillaries. However, when a nodular structure (the space between spheres) exists, a Kozeny-Carman equation is usually applied:

$$J = \frac{\varepsilon^3}{K\mu S^2(1-\varepsilon)^2} \frac{\Delta P}{\Delta z} \qquad (3.30)$$

where:
- S is the internal surface area
- ε is the volume fraction of the pores
- τ is the pore tortuosity
- μ is the viscosity
- K is the Kozeny-Carman constant.

Ultrafiltration (UF)

Ultrafiltration (UF) is the most common membrane process used in the food industry, and it involves the use membrane with a pore size ranging between 0.01–0.2 μm. Ultrafiltration is not as fine a filtration process as nanofiltration, but it also does not require as much energy to perform the separation. Applications of ultrafiltration in food processing can mostly likely be found in situations that call for separating one or more desirable components from a liquid mixture.

Ultrafiltration is capable of concentrating bacteria, some proteins, some fats, some colloidal minerals and constituents that have a larger molecular weight of greater than 10 kDa, but it is typically not effective at separating organic streams (Rosenberg, 1995). In UF, the chemical nature of membrane materials has only a small effect upon the separation, since ultrafiltration separation, like microfiltration, is based upon sieving mechanisms. Thus, ultrafiltration is only partially dependent upon the charge of the particle and is much more concerned with its size. The mass transfer equations for UF are similar to those for MF.

Reverse osmosis (RO)

Reverse osmosis (RO), also known as hyperfiltration, is the finest filtration known. This process will allow the removal of particles as small as ions from a solution. Reverse osmosis is used to purify water and to remove salts and other impurities in order to improve the color, taste or properties of the fluid. It can be used to purify wastewater streams that need additional treatment to remove water, which the reverse osmosis membrane will allow to pass through while rejecting other ions and colloids from passing.

The most common use for reverse osmosis is in purifying water. It is used to produce water that meets the most demanding specifications that are currently in place. It uses a membrane that is semi-permeable, allowing the fluid that is being purified to pass through it, while rejecting the contaminants that remain. Most reverse osmosis technology uses a process known as cross-flow to allow the membrane to continually clean itself. As some of the fluid passes through the membrane, the rest continues downstream, sweeping the rejected species away from the membrane.

The process of reverse osmosis requires a driving force to push the fluid through the membrane, and the most common force is pressure from a pump. A reverse osmosis process involves pressures 5–10 times higher than those used in ultrafiltration. As the concentration of the fluid being rejected increases, the driving force required for continuing the process of concentrating the fluid increases.

Reverse osmosis is capable of rejecting bacteria, salts, sugars, proteins, particles, fats, and other constituents that have a molecular weight of greater than 0.15–0.25 kDa. The separation of ions is aided by charged particles, which means that dissolved ions that carry a charge, such as salts, are more likely to be rejected by the membrane than those which are not charged, such as organics. The larger the charge and the larger the particle, the more likely the particle will be rejected. The transport mechanism of RO, as well as of that of nanofiltration, is now believed to be the solution diffusion mechanism. The evaluation of RO performance can be conducted with Equations (3.24), (3.25), (3.26), and (3.27).

Nanofiltration (NF)

Nanofiltration (NF) is a form of filtration that uses membranes to separate different fluids or ions preferentially. Nanofiltration is not as fine a filtration process as reverse osmosis, but it also does not require as much energy to perform the separation. It also uses a membrane that is partially permeable to perform the separation, although the pores of the membrane are typically much larger than those used in reverse osmosis. Nanofiltration is most commonly used to separate a solution that has a mixture of some desirable components and some that are not desirable. An example of this is the concentration of corn syrup. The nanofiltration membrane will allow the water to pass through the membrane, while holding the sugar back, concentrating the solution. As the concentration of the fluid being rejected increases, the driving force required to continue concentrating the fluid increases.

Nanofiltration is capable of concentrating sugars, divalent salts, bacteria, proteins, particles, fats, and other constituents that have a molecular weight greater than 1 kDa. Like reverse osmosis, it is affected by the charge of the particles being rejected, i.e., particles with larger charges are more likely to be rejected than others. The mass transport mechanism of NF and the membrane material used are quite comparable to those of RO. In some cases, NF is grouped into reverse osmosis processes.

3.10.2 Membrane separations by electrical potential difference: electrodialysis (ED)

Electrodialysis (ED) is an electrically driven membrane separation process that is capable of separating, concentrating, and purifying selected ions from aqueous solutions (as well as some organic solvents). The process is based on the property of ion exchange membranes to reject anions or cations selectively. If membranes are more permeable to cations than to anions or vice versa, the concentration of ionic solutions increases or decreases, so that concentration or depletion of electrolyte solutions is possible. In electrodialysis, only ionic species are transferred directly; thus, removal of ionic species from non-ionic products can be accomplished so that purification is also possible.

Electrodialysis Reversal (EDR) is an electrodialysis process in which the polarity of the electrodes is reversed on a prescribed time cycle, thus reversing the direction of ion movement in a membrane stack. The advantage of EDR is that it mitigates some of the concentration polarization and membrane fouling problems (Davis, 1990). The largest application of ED is the production of drinkable water from brackish water. Electrodialysis can remove salts from food, dairy, and other products, as well as concentrate salts, acids or bases. It also finds applications in wine and juice stabilization and in removing unwanted total dissolved solids that can build up in product streams (López-Leiva, 1988; Davis, 1990).

Faraday's Law supplies the basis to model ion transport and affirms that the total current in an electrolytic cell is equal to the sum of the electricity conveyed by each ion species:

$$I = f \sum (J_i Z_i) = fQ \sum \left(Z_i \frac{\Delta C_i}{e_i} \right) \qquad (3.31)$$

where:
   I is the current density
   f is Faraday's constant
   Q is the flow rate
   ΔC_i is the concentration difference
   J_i is the molar flux
   e_i is the current efficiency
   Z_i is the valence of ion i.

Concentration polarization also severely affects the current density, and the diffusive flux (the current density) through the concentration gradient over the boundary layer for a univalent ionic solution ($Z = 1$) is:

$$I = \frac{Df(C_b - C_m)}{\delta_c(t_m - t_{bl})} \quad (3.32)$$

where:
  D is the diffusivity
  C_m and C_b are concentrations at the membrane surface and in the bulk, respectively
  δ_c is the thickness of the concentration boundary layer
  t_m and t_{bl} are the transport numbers of the ion in the membrane and in the solution, respectively.

The transport number of the ion is defined as:

$$t_i = \frac{(J_i Z_i)}{\Sigma(J_i Z_i)} \quad (3.33)$$

3.10.3 Membrane separations by partial vapor pressure gradient

Pervaporation (PV)

Pervaporation (PV) is the separation of liquid mixtures by partial vaporization through a dense permselective membrane. Unlike the other membrane processes, a phase change occurs when the permeate changes from liquid to vapor during its transport through the membrane. PV is, in fact, an enrichment technique similar to distillation; however, unlike distillation, PV is not limited by the vapor–liquid equilibrium. PV has been commercially applied to the separation of azeotropic mixtures (dehydration of alcohol).

The heart of the PV is a non-porous membrane, which either exhibits a high permeation rate for water but does not permeate organics, or vice versa. A gradient in the chemical potential of the substances on the feed side and the permeate side creates the driving force for the process, which can be represented by partial vapor pressures on both sides of the membrane. The driving force is kept at a maximum by applying low pressure (vacuum or sweep gas) to the permeate side of the membrane, combined with immediate condensation of permeated vapors.

Pervaporation processes have found uses in the chemical industry to break azeotropic water/alcohol mixtures and to perform separations that are highly energy-intensive when distillation is used. Over the decades, a growing amount of attention has been paid to the application of pervaporation to solve environmental problems involving contaminated water.

The performance of pervaporation is commonly evaluated by two experimental parameters, namely the permeation flux and the selectivity. The performance of a pervaporation process is assessed by the flux of the permeating species and the selectivity of the species:

$$J_i = k^{ov}_i \rho'[C_i^L - C_i^v] \quad (3.34)$$

where $k^{ov}_i, \rho', C_i^L, C_i^v$ are the overall mass transfer rate constant, molar density of feed, bulk liquid phase concentration (mole fraction), and bulk vapor phase concentration, respectively, for component i.

The most commonly used selectivity parameter is separation factor, as shown in Equation (3.23). Sometimes, however, the enrichment factor, β_i, is used as an indication of the separation selectivity for component i:

$$\beta_i = \frac{(C_i)^v}{(C_i)^L} \quad (3.35)$$

As the concentration of component i is reduced, the concentration of component j will approach 1. The separation factor will therefore be close to the value of the enrichment factor, β_i, for dilute solutions:

$$\alpha_{ij} \approx \beta_i \quad (3.36)$$

The pervaporative transport process follows the solution-diffusion model, which is also the transport mechanism of RO and NF and consists of the following steps:

- Diffusion through the liquid boundary layer next to the feed side of the membrane.
- Selective partitioning of molecules of components into the membrane.
- Selective transport (diffusion) through the membrane matrix.
- Desorption into vapor phase on the permeate side.
- Diffusion away from the membrane and into the vapor boundary layer on the permeate side of the membrane.

Often, each step can be modeled with different approaches and fundamental assumptions. However, as with all mass transfer operations, the slowest step in this sequence will limit the overall rate of mass transfer and will be the center of research focus. Naturally, these steps are conveniently expressed in the form of the resistance-in-series model, which is expressed with mathematical symbols as:

$$\frac{1}{k^{ov}} = \frac{1}{k^{bl}} + \frac{1}{k^m} + \frac{1}{k^v} \quad (3.37)$$

The ks appearing in the equation are mass transfer coefficients, and their reciprocals represent the mass transfer resistance at each step. For many pervaporation processes, the mass transfer resistance in the vapor boundary layer tends to be small enough to be ignored. This leaves only the liquid boundary layer ($1/k^{bl}$) and membrane ($1/k^m$) resistances to deal with. k^m is strongly determined by polymer properties, the thickness of the membrane, and chemical structures of the components in the liquid.

In situations where hydrophobic aroma compounds are being removed from orange juice by pervaporation, the mass transfer rate is often limited by diffusion of the compound in the liquid boundary layer, i.e., $k^{ov} \approx k^{bl}$. This situation arises because nowadays membrane can be made in such a way that it provides minimal mass transfer resistance to the aroma compounds, almost to the point of being absent. This situation manifests itself as a phenomenon called "concentration polarization", i.e., the steep discrepancy in aroma compound concentrations between the bulk and the membrane surface. When concentration polarization is severe, the function of membrane is to minimize the solvent flux, thereby maximizing the selectivity of the intended separation.

The analysis of the liquid boundary layer mass transfer resistance is very important to the process designers and the operators alike. One common approach to the analysis is to find out the correlation between mass transfer coefficient and process parameters. It is recommended that the boundary-layer theory would have to be adopted to provide more robust analysis that has broader application and scalability. However, in reality, it is exceedingly difficult to do this. Instead, semi-empirical correlations that have the form of Sherwood number correlations, as shown in Equation (3.28), are commonly employed. Among these correlations is the frequently cited Lévêque's correlation:

$$k^{bl} = 1.6 \frac{D}{d_h} Re^{1/3} Sc^{1/3} \left(\frac{d^h}{L}\right)^{1/3} \quad (3.38)$$

The only new parameter in this correlation is L, the length of the membrane channel. This correlation indicates that the mass transfer coefficient is mainly dependent upon flow conditions on the feed side and the shape and dimensions of the module. Temperature has a substantial impact upon the mass transfer coefficient through the diffusivity of the solute and viscosity. Nevertheless, temperature is not a commonly manipulated parameter, due to the issues of membrane stability and the vapor pressure of the volatile solute, since the feed has to be kept in the liquid phase. Flow velocity (flow rate) is the parameter that can be adjusted to minimize the liquid boundary layer resistance for a fixed membrane module (configuration).

Membrane distillation (MD)

Membrane distillation (MD) is a type of low-temperature, reduced pressure distillation using porous hydrophobic (water-hating) polymer materials. It is a process that separates two aqueous solutions at different temperatures, and it has been developed for the production of high-purity water and for the separation of volatile solvents, such as acetone and ethanol. MD can achieve higher concentrations than RO.

In MD, the membrane must be hydrophobic and microporous. The hydrophobic nature of the material prevents the membrane from being wet by the liquid feed and, hence, liquid penetration and transport across the membrane is avoided, provided the feed side pressure does not exceed the minimum entry pressure for the pore size distribution of the membrane. The driving force of MD is temperature gradient, and the two different temperatures produce two different partial vapor pressures at the solution-membrane interface, which propels consequent penetration of the vapor through the pores of the membrane. The vapor is condensed on the chilled wall by cooling water, producing a distillate. This process usually takes place at atmospheric pressure and a temperature that may be much lower than the boiling point of the liquids (e.g., solvents). It is commonly observed that the effect of the osmotic pressure from the permeate to the feed solution will be prominent when the high solute concentrations of feed liquids are processed.

A variation of MD, sometimes called "low pressure membrane distillation" or "osmotic distillation", uses an auxiliary device to condense the vapor coming out of the membrane. The driving force for vapor transport, in this case, is the pressure differential. Alternatively, the auxiliary cooling device can be replaced by using an inert sweep gas or absorbing strip liquid to remove the vapor permeate and maintain the pressure differential. The membrane distillation is very similar to a single-stage distillation and, thus, is unable to achieve a high separation factor. The primary advantage of MD is the high surface area/volume ratio available and, thus, high permeation rates. Most food applications of MD are focused on dehydration of liquid foods.

The performance of an MD can be evaluated by a phenomenological equation:

$$J = F\Delta P \qquad (3.39)$$

where the flux is related to two parameters; one is pressure difference and the other proportionality factor ("membrane permeability") F. ΔP is mainly determined by the temperature difference ΔT, which can be related to the Clausius-Claperyron relationship:

$$\ln P = \frac{\Delta H_{vap}}{RT} + c \qquad (3.40)$$

where:

H_{vap} is the enthalpy of the vapor of permeating species
T is the temperature
C is the constant.

MD sometimes experiences a temperature polarization phenomenon, due to the difference in heat transfer rate between the heat conduction in the membrane and the heat transfer in the bulk fluid.

3.10.4 Membrane contactor (MC)

Membrane contactors (MCs) are a motley group of several membrane processes that primarily use the membranes as mass transfer barriers for certain materials and interfaces between two phases. The driving force in the process is typically the difference in either vapor pressures or the osmotic pressures across the membrane barrier. The relevant membrane contactors for fruit juice processing are direct osmosis, membrane distillation, and osmotic distillation.

Direct osmosis originates from an old practice for juice concentration and is caused by the difference of osmotic pressures between dilute juice on the upstream side and brine on the other side. The disadvantages of DO are the high costs and low permeation rates, even through DO concentration can reach concentrations greater than those achieved by RO.

Membrane Distillation (MD) utilizes the vapor pressure difference across the membrane resulting from temperature difference to drive the solvent molecules across the microporous hydrophobic membrane to condense at the cold side of the membrane. Since temperature difference (and, subsequently, vapor pressure difference) drives MD, so an increase in temperature on one side of the membrane would increase permeation rate. However, temperature polarization becomes prominent at high temperatures and exerts a negative effect on permeation of MD. The other problem with MD operations is that the process is limited in terms of operating temperature, due to concerns about thermal damage to the flavor compounds in fruit juices.

The above limitation with MD makes osmotic distillation (a.k.a. osmotic evaporation or isothermal membrane distillation) a better choice. In osmotic distillation (OD), a liquid mixture containing volatile component is contacted with a microporous, non-wettable membrane whose opposite surface is exposed to another liquid phase, where the mass transfer takes place across the membrane. This technology can selectively remove the water from liquid wastes under atmospheric pressure and at room temperature. Like other membrane contactors, OD suffers from high costs and low permeation rate.

3.10.5 Design considerations

In many cases, it is still true to say that the design of a membrane process or/and the selection of a membrane module/material for desired separations remains a mixture of art and science, in which knowledge, experience, and science all play important roles. More often than not, there is no "right answer" in the absolute sense, as more than one solution is often both technically and economically viable. However, a careful evaluation at the outset of as many as possible of the factors influencing the choices will help narrow down the items on the list.

When contemplating the use of any particular membrane process for the separation of components in a liquid food stream, after the initial assessment, several process issues must be evaluated. The first step in doing so is to draw up the detailed requirements for the process. Accurate qualitative and, where possible, quantitative information on the following aspects should therefore be specified:

- the components and range of concentrations in the feed;
- the intended use or fate of the treated feed liquids (i.e., final products, further processing, etc.);
- the intended use of or fate of the permeate (i.e., disposal, reuse, further processing, etc.);
- permeation flux;
- the minimum properties of the treated food fluids and permeate that will make the intended use or fate possible;
- membrane transport mechanism;
- cost-effective and environmentally friendly alternative solution.

Any one or more of the above factors may, depending upon circumstances, influence the design of a membrane process. In the case of food-related wastes and by-products, it is normal for economical considerations to override operational simplicity. This is particularly true for aroma compound recovery in orange juice production effluents using pervaporation, in which the preservation of the permeate is the key to exerting a major influence on the ultimate cost of treating the wastewater stream. In other processing applications, a membrane process can be an important intermediate operation that is vital to subsequent processing operations.

The next issue to be addressed is whether the membrane processes are actually capable of separating the components from liquid foods. The answer to this question for pressure-driven membrane processes such as MC, UF, NF, and RO is generally affirmative, provided that appropriate membranes (pore size, and for NF and RO membrane properties such as charge, hydrophilic tendency) are used. For pervaporation, the answer is more complicated and conditional. It is well documented that PV works well when the compound to be removed has a high vapor pressure relative to the background material and a low solubility in the background material. In dilute aqueous solutions, such as aromas in orange juice, it is generally the Henry's constant that determines whether an aroma compound can effectively separated by pervaporation. The Henry's constant represents the vapor-liquid partitioning of organic compounds in an aqueous system. The general rule of thumb is: the more dissimilar the components, the easier it will be to separate them.

Once the process designer has determined that a particular membrane process will, theoretically, work, he or she will need to ask these subsequent questions:

- Do membrane materials exist which will do the job?
- Is this membrane material available in a membrane module?

The answer to the first question is usually positive. A great deal of research has been performed on many membrane materials and feed mixtures. In addition, a wide array of membrane materials is available that may achieve the desired separation, but which have not been tested in a membrane mode of interest.

Another variable in the selection of membrane materials is whether a single-layer membrane or a multi-layer membrane is to be used. Membranes used in an MF are normally single-layer and isotropic, while membranes in other pressure-driven membrane filtrations and pervaporation are composed of composite or multi-layer, non-homogenous materials. This is because a membrane with desired selectivity may require a significant thickness to deliver the desired physical properties such as burst pressure. However, improving membrane mechanical stability by increasing membrane thickness would inadvertently reduce permeation flux. To get around this problem, a composite

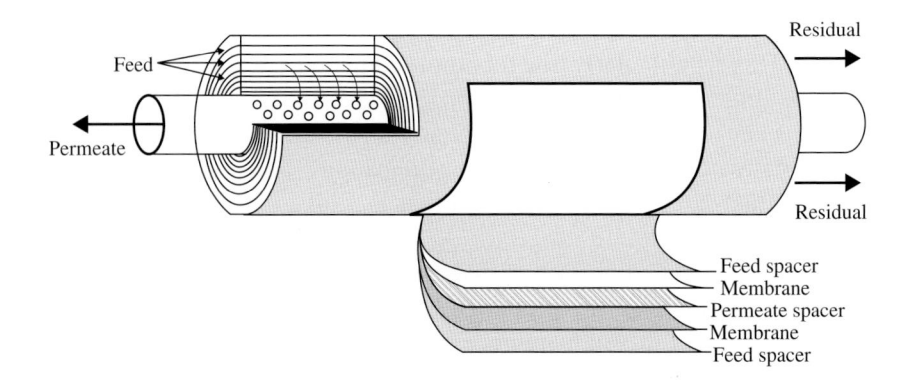

Figure 3.14 Schematic illustration of a spiral wound module (Courtesy of Dr. Leland Vane of USEPA) (see plate section for color version)

or inhomogeneous membrane is employed, where a thick layer of polymer material with large pore size supports a top thin layer of active membrane.

The second question concerns the issue of commercial availability of membrane configurations or membrane modules for particular membrane materials. A module is the smallest unit into which the membrane material is packed. The reason for using modules is because, although polymer membranes are made in two basic physical forms – flat sheet and tubular – many practical membrane systems that need large membrane areas can only be accommodated in membrane modules. For pressure-driven membrane processes, membrane distillation and pervaporation, there are four primary configurations (modules), each with inherent advantages and weaknesses. These four are: spiral wound; hollow fiber; plate and frame; and tubular.

3.10.6 Membrane modules

Spiral Wound Module

A spiral wound module is a logical step from a flat sheet membrane. A flat membrane envelope or set of envelopes is rolled into a cylinder, as shown in Figure 3.14.

The envelope is constructed from two sheets of membrane, sealed on three edges. The inside of the envelope is the permeate side of the membrane, and a thin porous spacer inside the envelope keeps the two sheets separated. The open end of the envelope is sealed to a perforated tube (the permeate tube) with glue, so that the permeate can pass through the perforations. For pervaporation, this is also the place to which the vacuum or sweep gas is applied. Another spacer is laid on top of the envelope before it is rolled, creating the flow path for the feed liquid. This feed spacer generates turbulence due to the undulating flow path that disrupts the liquid boundary layer, thereby enhancing the feed side mass transfer rate.

It is the fact that the envelopes and spacers are wrapped around the permeate tube that gives the module its name – spiral wound module. The spiral wrapped envelopes and spacers are then wrapped again with tape or glass, or a net-like sieve, before being fitted into a pressure vessel. In this way, a reasonably sized membrane area can be housed in a convenient module, resulting in a very high surface area-to-volume ratio.

One noticeable drawback lies in the permeate path length. A permeating component that enters the permeate envelope farthest from the permeate tube must spiral inward several feet. Depending upon the path length, permeate spacer design, gel layer, and permeate flux, significant permeate side pressure drops can be encountered. The other disadvantage of this module is that it is a poor choice for treating fluids containing particulate matters.

Hollow fiber module

In a hollow fiber configuration, small-diameter polymer tubes are bundled together to form a hollow fiber module like a shell and tube heat exchanger (Figure 3.15).

These modules can be configured for liquid flow on the tube side, or lumen side (inside the hollow fibers), or vice versa. These tubes have diameters on the order of 100 microns, giving them a very high surface area-to-module volume ratio. This makes it possible to construct compact modules with high surface areas. The drawback is that the liquid flow inside the hollow fibers is normally within the range of the laminar flow regime due to its low hydraulic diameter. The consequence of prevalent laminar flows is high mass transfer resistance on

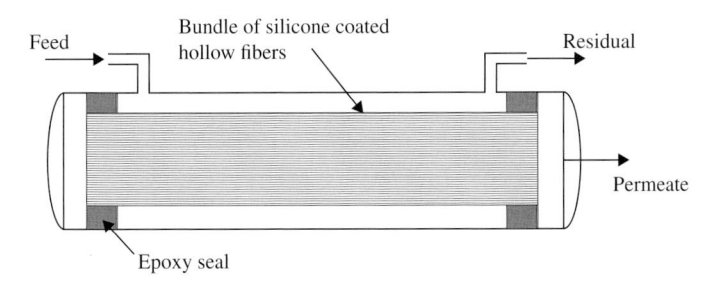

Figure 3.15 Schematic illustration of a hollow fiber module (Courtesy of Dr. Leland Vane of US EPA)

the liquid feed side. However, because of the laminar flow regime, the modeling of mass transfer in a hollow fiber module is relatively easy, and the scale-up behavior is more predictable than that in other modules. One noticeable problem with a hollow fiber module, though, is that a whole unit has to be replaced if failure occurs.

Plate and frame module

Plate-and-frame configuration is a migration from filtration technology. It is formed by the layering of flat sheets of membrane between spacers. The feed and permeate channels are isolated from one another using flat membranes and rigid frames (Figure 3.16).

This configuration was an early favorite, and it is a natural scale-up from bench-scale laboratory membrane cells that have one feed chamber and one permeate chamber separated by a flat sheet of membrane. A single plate and frame unit can be used to test different membranes by swapping out the flat sheets of membrane. Furthermore, it allows for the use for membrane materials that cannot be conveniently produced as hollow fibers or spiral wound elements. The

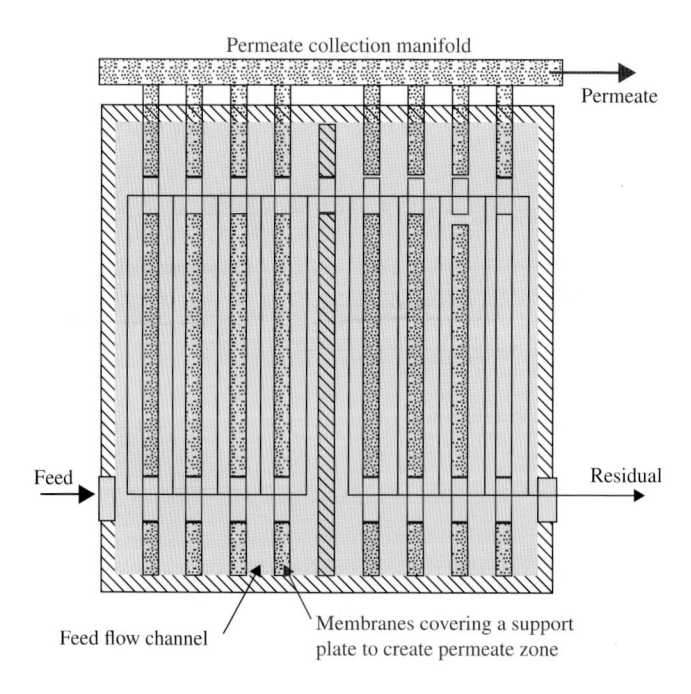

Figure 3.16 A schematic illustration of a plate and frame module (Courtesy of Dr. Leland Vane of US EPA) (see plate section for color version)

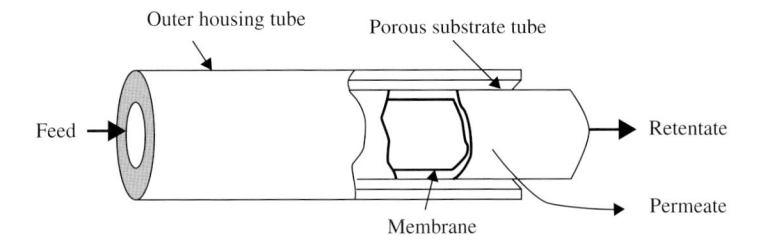

Figure 3.17 Schematic illustration of a tubular module

disadvantages are that the ratio of membrane area-to-module volume is low compared to spiral wound or hollow fiber modules, dismounting is time-consuming and labor-intensive, and there are higher capital costs associated with the frame structures.

Tubular module

Polymeric tubular membranes are usually made by casting a membrane onto the inside of a pre-formed tube, which is referred to as the substrate tube. The tube is generally made from one or two piles of non-woven fabric, such as polyester or polypropylene. The diameters of tubes range from 5–25 mm (Figure 3.17).

A popular method of construction of these tubes is a helically wound tape that is welded at the edges. The advantage of the tubular membrane is its mechanical strength if the membrane is supported by porous stainless steel or plastic tubes. Tubular arrangements often provide good control of flow to the operators, and are easy to clean. Additionally, this is the only membrane format for inorganic membranes, particularly ceramics. The disadvantage of this type of modules is mainly higher costs in investment and operation. The arrangement of tubular membranes in a housing vessel is similar to that of hollow fiber element. Tubular membranes are sometimes arranged helically to enhance mass transfer by creating a second flow (Dean vortex) inside the substrate tube (Moulin *et al.*, 1999).

Although the specification for the process is the most critical issue in the process design of membrane systems, certain auxiliary steps must also be considered in the operation of a membrane system. For example, temperature, pH limits and tolerance to certain chemicals – particularly cleaning agents such as alkalis and detergents – should be considered before a process is put on line. These cleaning chemicals, as well as seals and glues used in the membrane modules, have to be approved by FDA or other regulatory agencies for used in food processing – an aspect of process design that is often neglected by some designers.

3.10.7 Whose fault? membrane fouling

As described in the previous sections, concentration polarization phenomena in membrane processes can cause noticeable decline in the membrane's performance. In membrane filtration processes such as microfiltration and ultrafiltration, concentration polarization phenomena are always accompanied by the formation of a gel layer that is either irreversible or reversible. The cause of gel layer formation is thought to be the result of the rapid accumulation of retained solutes near the membrane surface, to the point that the concentration of macromolecule solute reaches the gel-forming concentration. High retention of solutes near the membrane surface inevitably also leads to concentration polarization and, as a result, the performances of membrane filtration processes (pressure-driven processes) suffer.

The version of concentration polarization used in pervaporation is slightly different from its kindred in membrane filtration, as stated in the previous section. Concentration polarization also has a negatively effect on the performance of an electrodialysis process. For membrane distillation, temperature polarization is the main culprit for the decline in the process performance.

Membrane fouling is suspected if the membrane flux is continuously declining after a period of operation. This is usually an irreversible, partially concentration-dependent and time-dependent phenomenon, which distinguishes it from concentration polarization. The identification of membrane fouling is imprecise and is often based upon the operator's experience, performing fouling tests with membrane filtration index apparatus, and membrane vendor's recommendations. Membrane fouling is intimately related to concentration polarization, but the two are not exactly interchangeable in our description of membrane performance deterioration.

We now know that all membrane filtration processes experience some degree of concentration polarization, but fouling occurs mainly in microfiltration and ultrafiltration. Relatively large pores in these membranes are implicitly vulnerable to fouling agents such as organic and inorganic precipitates, as well as fine particulate matters that could lodge in these pores or deposit irreversibly on the membrane surface. The exact cause of membrane fouling is very complex and is, therefore, difficult to depict in full confidence with available theoretical understandings. Even for a known solution, fouling is influenced by a number of chemical and physical parameters, such as concentration, temperature, pH, ionic strength, and specific interactions (e.g., hydrogen bonding, dipole-dipole interactions).

Membrane fouling in membrane filtration processes, like concentration polarization, is unavoidable, and this is particularly true for protein concentration or fractionation. However, certain steps that will greatly reduce the severity of membrane fouling can still be achieved. One effective way to reduce membrane fouling is to provide pre-treatment to the feed liquids. Some simple adjustments, such as varying pH values and using hydrophilic membrane materials, can also do wonders in protein concentration operations. There is persistent interest around the world in modifying membrane properties to minimize the tendency towards membrane fouling. Since it is intimately associated with concentration polarization phenomenon, any action taken to minimize concentration polarization will also benefit the fight against membrane fouling.

Unfortunately, no matter how much effort put forward fighting against membrane fouling, it will eventually occur. The only solution by then is employing a cleaning regimen. The frequency of cleaning necessary depends on many factors and should be considered as a part of the process optimization exercise. There are three basic types of cleaning methods currently used: hydraulic flushing (back-flushing); mechanical cleaning (only in tubular systems) with sponge balls; and chemical washing. When using chemicals to perform de-fouling, precautions must be observed, since many polymeric membrane materials are susceptible to chlorine, high pH solutions, organic solvents, and a host of other chemicals.

3.11 Ion exchange

Ion exchange is a process in which ions of a particular species in solution are replaced by ions with a similar charge, but of different species, attached to an insoluble resin. In essence, ion exchange is a sorption process and can also be considered as a reversible chemical reaction. The common applications of ion exchange are water softening (a.k.a. removal of "hardness" ions such as Ca^{2+} and Mg^{2+}) and nitrate removal in advanced wastewater treatment operations. These ion exchange resins are either naturally occurring inorganic zeolites or synthetically produced organic resins. The synthetic organic resins are the predominant type used today, because their characteristics can be tailored to specific applications.

An organic ion exchange resin consists of an organic or inorganic network structure, with attached functional groups that can exchange their mobile ions for ions of similar charge from the surrounding medium. Each resin has a distinct number of mobile ion sites that set the maximum quantity of exchanges per unit of resin. The resins are called cationic if they exchange positive ions, or anionic if they exchange negative ions. Cation exchange resins have acidic functional groups such as sulfonic, whereas anion exchange resins are often classified by the nature of the functional group as strong acid, weak acid, strong base, and weak base. The strength of the acidic or basic character depends on the degree of ionization of the functional groups, similar to the situation with soluble acids or bases. Accordingly, a resin with sulfonic acid groups would act as a strong cation exchange resin.

Ion exchange reactions are stoichiometric and reversible, and in that way they are similar to other solution phase reactions. For example:

$$MgSO_4 + Ca(OH)_2 \rightarrow Mg(OH)_2 + CaSO_4 \qquad (3.41)$$

In this reaction, the magnesium ions from magnesium sulfate ($MgSO_4$) are exchanged for calcium ions from the calcium hydroxide $Ca(OH)_2$ molecule. Similarly, a resin with hydrogen ions available for exchange will exchange those ions for magnesium ions from solution. The reaction can be written as follows:

$$2(R\text{-}SO_3H) + MgSO_4 \rightarrow 2(R\text{-}SO_3)_2Mg + H_2SO_4 \qquad (3.42)$$

where:
    R indicates the organic portion of the resin
    SO_3 is the immobile portion of the ion active group.

Two resin sites are needed for magnesium ions with a 2+ valence (Mg^{2+}).

As stated previously, the ion exchange reaction is reversible. The degree to which the reaction proceeds to the right will depend on the resin's preference, or selectivity, for magnesium ions, compared with its preference for hydrogen ions. The selectivity of a resin for a given ion is measured by the selectivity coefficient, K, which, in its simplest form for the reaction,

$$R - A^+ + B^+ \rightarrow R - B^+ + A^+ \qquad (3.43)$$

is expressed as $K =$ (concentration of B^+ in resin/concentration of A^+ in resin) $\times$ (concentration of A^+ in solution/concentration of B^+ in solution).

The selectivity coefficient expresses the relative distribution of the ions when a resin in the A^+ form is placed in a solution containing B^+ ions. Table 3.1 shows the selectivities of strong acid and strong base ion exchange resins for various ionic compounds.

It should be pointed out that the selectivity coefficient is not constant, but varies with changes in solution conditions. It does provide a means of determining what to expect when various ions are involved. As indicated in Table 3.1, strong acid resins have a preference for magnesium over hydrogen. Despite this preference, the resin can be converted back to the hydrogen form by contact with a concentrated solution of sulfuric acid (H_2SO_4):

$$(R\text{-}SO_4)_2Mg + H_2SO_4 \rightarrow 2(R\text{-}SO_3H) + MgSO_4 \qquad (3.44)$$

Table 3.1 Selectivity of ion exchange resins for some ions in order of decreasing preference

Strong acid cation exchanger	Strong base anion exchanger
Barium	Iodide
Calcium	Nitrate
Copper	Bisulfite
Zinc	Chloride
Magnesium	Cyanide
Potassium	Bicarbonate
Ammonia	Hydroxide
Sodium	Fluoride
Hydrogen	Sulfate

(Source: Weber, 1972).

This step is known as regeneration. In general terms, the higher the preference a resin exhibits for a particular ion, the greater the exchange efficiency in terms of resin capacity for removal of that ion from solution. Greater preference for a particular ion, however, will result in increased consumption of chemicals for regeneration.

Ion exchange resins are classified as either cation exchangers, which have positively charged mobile ions available for exchange, or anion exchangers, whose exchangeable ions are negatively charged. Both anion and cation resins are produced from the same basic organic polymers, differing in the ionizable group attached to the hydrocarbon network. It is this functional group that determines the chemical behavior of the resin. Resins can be broadly classified as strong or weak acid cation exchangers, or strong or weak base anion exchangers.

3.11.1 Strong acid cation resins

Strong acid resins are so named because their chemical behavior is similar to that of a strong acid. The resins are highly ionized in both the acid ($R-SO_3H$) and salt ($R-SO_3Na$) form. They can convert a metal salt to the corresponding acid by the reaction:

$$2(R-SO_3H) + MgCl_2 \rightarrow (R-SO_3)_2Mg + 2HCl \qquad (3.45)$$

The hydrogen and sodium forms of strong acid resins are highly dissociated, and the exchangeable Na^+ and H^+ are readily available for exchange over the entire pH range. Consequently, the exchange capacity of strong acid resins is independent of solution pH. These resins are used in the hydrogen form for complete deionization, or in the sodium form for water softening (calcium and magnesium removal). After exhaustion, the resin is converted back to the hydrogen form (regenerated) by contact with a strong acid solution, or the resin can be converted to the sodium form using a sodium chloride solution. Hydrochloric acid (HCl) regeneration would result in a concentrated magnesium chloride ($MgCl_2$) solution.

3.11.2 Weak acid cation resins

In a weak acid resin, the ionizable group is a carboxylic acid (COOH), as opposed to the sulfonic acid group (SO_3H) used in strong acid resins. These resins behave similarly to weak organic acids that are weakly dissociated.

Weak acid resins exhibit a much higher affinity for hydrogen ions than do strong acid resins. This characteristic allows for regeneration to the hydrogen form with significantly less acid than is required for strong acid resins. Almost complete regeneration can be accomplished with stoichiometric amounts of acid. The degree of dissociation of a weak acid resin is strongly influenced by the solution pH. Consequently, resin capacity depends in part on solution pH.

3.11.3 Strong base anion resins

Like strong acid resins, strong base resins are highly ionized and can be used over the entire pH range. These resins are used in the hydroxide (OH) form for water deionization. They will react with anions in solution and can convert an acid solution to pure water:

$$R\text{-}NH_3OH + HCl \rightarrow R\text{-}NH_3Cl + HOH \qquad (3.46)$$

Regeneration with concentrated sodium hydroxide (NaOH) converts the exhausted resin to the hydroxide form.

3.11.4 Weak base anion resins

Weak base resins are like weak acid resins, in that the degree of ionization is strongly influenced by pH. Consequently, weak base resins exhibit minimum exchange capacity above a pH of 7.0. These resins merely sorb strong acids; they cannot split salts.

Table 3.2 A commercial ion exchange resin property sheet (Courtesy of DOW Chemical Company)

Commercial name Type	DOWEX® 1 X ... Strongly basic anion exchanger	DOWEX® 50 WX ... Strongly acid cation exchanger
Functional group	trimethyl ammonium	Sulfonic acid
Cross linkage (% DVB)	2 or 8	2, 4 or 8
Ionic form as shipped	Cl^-	Na^+ (analytical grade) H^- (practical grade)
Shipping density (kg/L)	0.7	0.8
Volume change (%)	$Cl^- \geq OH^- \approx +20\%$	$Na^+ \geq H^+ \approx +8\%$
Effective working range (pH)	0 – 14	0 – 14
Selectivity for ions	$I^- > NO_3^- > Br^- > Cl^- >$ acetate$^- > OH^- > F^-$	$Ag^+ > Cs^+ > Rb^+ > K^+ > NH_4^+ > Na^+ > Li^+ > Ba^{2+} > Sn^{2+} > Ca^{2+} > Mg^{2+} > Be^{2+}$
Total exchange capacity (eq/L)	1.3	1.9
Thermal stability	OH^- form max. 50°C (122°F) Cl^- form max. 150°C (302°F)	Na^+ form max. 120°C (248°F) H^+ form max. 80°C (176°F)
Moisture (%)	39–80	40–82

DOWEX® is a registered trademark of Dow Chemical Company.

3.11.5 Evaluation of resins

Resin vendors usually provide detailed information regarding the properties of resins that they sell, as is shown in a typical ion exchange resin property sheet from DOW Chemical (Table 3.2). However, it is still sensible to evaluate the resins in service for any change in their capacity. The potential loss of active ion exchange sites due to reduction of cross-linking and other deleterious effects of long-term services is of particularly concern.

The common properties of resins for uses in wastewater treatment undergoing evaluations are:

- Dry weight capacity.
- Wet weight capacity.
- Wet volume capacity.
- Percentage of moisture content.

3.11.6 Ion exchange systems

In order to design an ion exchange system for removing ions from complex food and agricultural wastewater, several runs of a laboratory scale ion exchange column is necessary to develop system design criteria. Eckenfelder (1989) suggested an experimental procedure for conducting experiments on a lab-scale ion exchange column:

- Rinse the column for ten minutes with deionized water at a rate of 50 ml/min.
- Switch to a waste-containing solution passing through the column at the same flow rate as deionized water.
- Measure the initial volume of solution to be treated.
- Start the treatment cycle and develop the breakthrough curve until the ion concentration reaches the maximum effluent limit.
- Backwash to 25 & bed expansion for 5–10 minutes with distilled water.
- Regenerate at a flow rate of 6 ml/min using the concentration and volume recommended for the resin by the vendor, collect the spent regenerant and measure the recovered ions.
- Rinse the column with distilled water.

After several runs of the experiment, it is possible to select optimal operating conditions in terms of resin utilization and regenerant efficiency.

Most practical applications of ion exchange use fixed-bed column systems, the basic component of which is the resin column. Complete demineralization operations generally involve the wastewater passing first through a bed of strong acid resin to replace metal ions with hydrogen ions (thus lowering the pH), followed by a weakly basic anion exchanger, as shown in Figure 3.18, a schematic diagram of this two-stage type of arrangement.

Weak base resins are preferred over strong base resins because they require less regenerant chemical. A reaction between the resin in the free base form and HCl would proceed as follows:

$$R - NH_2 + HCl \rightarrow R - NH_3Cl \qquad (3.47)$$

The weak base resin does not have a hydroxide ion form, as does the strong base resin. Consequently, regeneration needs only to neutralize the absorbed acid; it does not need to provide hydroxide ions. Less expensive weakly basic reagents, such as ammonia (NH_3) or sodium carbonate can be employed.

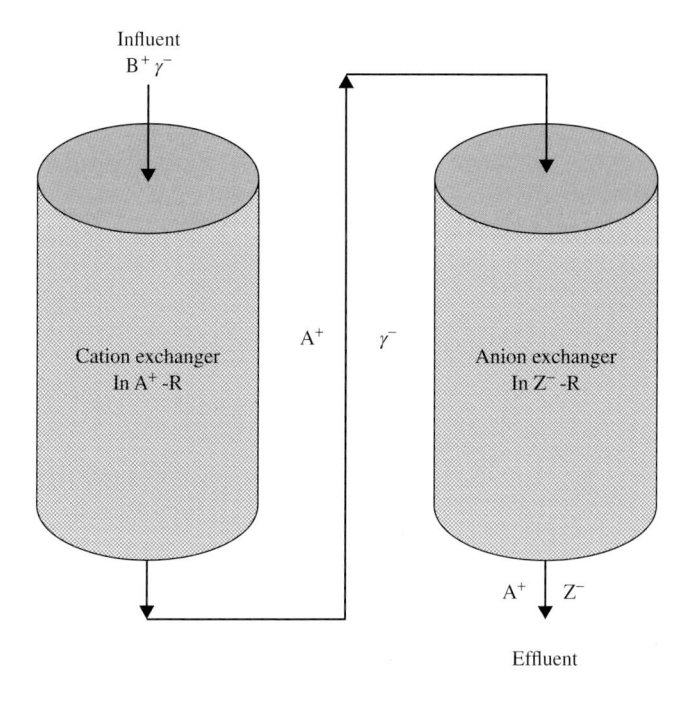

Figure 3.18 A schematic diagram of a two-stage ion exchange system

Ion exchanger systems used for wastewater treatment have been based on a process called DESAL (Downing et al., 1968), which utilizes a three-step operation:

- A weak base anion resin in the bicarbonate form, R- (NH)HCO3.
- A weak acid cation in the hydrogen form, R-COOH.
- A weak base anion resin in the free base form.

A schematic diagram of the DESAL process is shown in Figure 3.19.

3.12 Closing remarks

For organic-rich food and agricultural wastewater, biological treatment has unrivaled advantages. However, physicochemical processes are still important in treating this type of wastewater streams. First of all, physicochemical treatment plants have small footprints, which is important for densely populated areas. Second, physicochemical processes can be easily expanded as and when

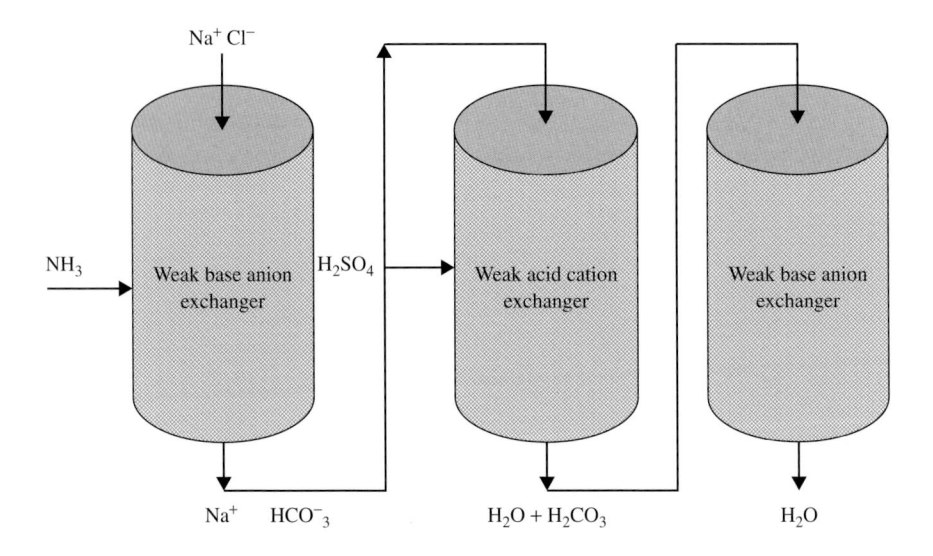

Figure 3.19 Schematic diagram of DESAL ion exchange system

required, for example if subsequent treatment using biological methods is planned. Third, the processes are often fast comparing to biological treatment, and may be included as an integral part of an overall wastewater management strategy if the influent streams are mixed with municipal or other industrial wastewater. Finally, certain pollutants in wastewater are not biodegradable, thus requiring physicochemical processes to remove them.

The disadvantages of physicochemical processes are well known: high operating and capital costs; relatively modest treatment performance; and larger sludge volume. It is no accident that physicochemical processes in practice are often interspersed with biological treatment processes to achieve optimal results. The ultimate choices of physicochemical processes for a given treatment task are largely dependent upon the deliberate consideration of technological and economical facts within the constraints of treatment requirements and regulatory compliance.

One of the most challenging aspects of treatment process design is the analysis and selection of the treatment processes capable of meeting the permit or recycling requirements. The methodology of process analysis that leads to process selection includes several evaluation steps, which vary greatly with the project and characteristics of wastewater. Nevertheless, any process analysis needs to consider several important factors: process applicability; applicable flow range and variation; reaction kinetics and reactor selection; performance; treatment residuals and odor; sludge treatment; chemicals/polymers requirements; and energy requirements.

Once process analysis is done, process selection or design commences. Several methods of process design or selection may be considered – process selection based on empirical relationship from experience or literature and process design based on kinetic analysis or modeling. Chapter 1 of this book provides the basic tools to assist the selection process.

3.13 Further reading

Cooney, D.O. (1998). *Adsorption Design for Wastewater Treatment*. CRC Press, Boca Raton, FL.
Drinan, Joanne E. (2000). *Water and Wastewater Treatment: A Guide for the Nonengineering Professionals*. CRC Press, Boca Raton, FL.
Droste, Ronald L. (1996). *Theory and Practice of Water and Wastewater Treatment*. Wiley, New York, NY.
Eckenfelder, W.W., Bowers, A.R. & Roth, J.A. (1996). *Chemical Oxidation: Technology for the Nineties*, Volume IV. CRC Press, Boca Raton, FL.
Hahn, H., Hoffman, E. & Odegaard, H. (2002). *Chemical Water and Wastewater Treatment VII: (Gothenburg Symposia)*. IWA Publishing, London, UK.
Lin, S.D. & Lee. C.C. (2001). *Water and Wastewater Calculations Manual*. McGraw-Hill Professional, New York, NY.
Metcalf & Eddy, Inc. (Tchobanoglous, G. & Burton, F.L). (1991). *Wastewater Engineering, Treatment, Disposal, and Reuse*, 3rd edition. McGraw-Hill, New York, NY.
Parsons, S. (2005). *Advanced Oxidation Processes for Water and Wastewater Treatment*. IWA Publishing, London, UK.
Qasim, S.R., (1998). *Wastewater Treatment Plants: Planning, Design, and Operation*, 2nd Edition. CRC Press, Boca Raton, FL.
Tang, Walter Z. (2003). *Physicochemical Treatment of Hazardous Wastes*. Lewis Publishers (CRC Press), Boca Raton, FL.
Visvanathan, C. & Ben Aim, R. (1989). *Water, Wastewater, and Sludge Filtration*. CRC Press. Boca Raton, FL.
Weber, Walter J., Jr. (1972). *Physicochemical Process: For Water Quality Control*. John Wiley & Sons, New York, NY.

3.14 References

Canale, R. P. and J.A. Borchardt. (1972) Sedimentation. In Walter J. Weber, Jr., *Physicochemical Process: For Water Quality Control*. New York, John Wiley & Sons.
Cheryan, M. (1986). *Ultrafiltration Handbook*. Technomic Publishing Co. Lancaster, PA.

Davis, T. A. (1990). Electrodialysis. In: Porter, M.C. (Ed.). *Handbook of Industrial Membrane Technology*. Noyes Publications, Park Ridge, NJ.

Downing, D.G., Kunin, R. & Pollio, F.X. (1968). *DESAL process–economic ion exchanger system for treating brackish and acid mine drainage waters and sewage waste effluents. Water–1968*. Chemical Engineering Progress Symposium Series, 64, 90. American Institute of Chemical Engineers, New York, NY.

Eckenfelder, W.W. (1989). *Industrial Water Pollution Control*, 2nd edition. New York, McGraw-Hill.

Glasgow, L.A. & Liu, S.X. (1995). Effects of Macromolecular Conformation upon Solid–Liquid Separation and Water Treatment Plant Residuals. *Environmental Technology* **16**, 915–927.

Liu, S.X. (1995). *The Essential Aspects of Floc Structure and Breakage*. Chemical Engineering Department, Kansas State University, Manhattan, KS.

López-Leiva, M. (1988). The Use of elctrodialysis in food processing, II. Review of practical applications. *Lebens-mittel Wissemschaft und Technologie* **21**, 177–182.

Metcalf & Eddy, Inc. (Tchobanoglous, G. & Burton, F.L). (1991). *Wastewater Engineering, Treatment, Disposal, and Reuse*, Third edition. McGraw-Hill, New York, NY.

Mulder, M. (1991). *Basic Principles of Membrane Technology*. Kluwer Academic Publishers, Dordrecht, Germany.

Moulin, P., Manno, P., Rouch, J.C., Serra, C., Clifton, M.J. & P. Aptel. (1999). Flux improvement by dean vortices: ultrafiltration of colloidal suspensions and macromolecular solutions. *Journal of Membrane Science* **156**, 109–130.

Rautenbach, R. & Albrecht, R. (1989). *Membrane Processes*. John Wiley & Sons, Chichester, UK.

Rosenberg, M. (1995). Current and future applications for membrane processes in the dairy industry. *Trends in Food Science and Technology* **6**, 12–19.

4

Biological wastewater treatment processes

4.1 Introduction

Biological wastewater treatment is often associated with secondary wastewater treatment, and its intended purpose is to treat the dissolved and colloidal organics after primary treatment. The goal of all biological wastewater treatment systems is to coagulate and remove or reduce the non-settling organic solids and the dissolved organic load from the effluents by using microbial communities to degrade the organic load through biochemical reactions. Biological wastewater treatment is generally a major part of secondary treatment design for wastewater and it is characterized by reduction of the oxygen demand of an influent wastewater to a given level of purification. The microorganisms responsible for reducing the organic materials and, consequently, the oxygen demand of incoming wastewater, can be classified as oxygen-aerobic (need oxygen for their metabolism), anaerobic (thrive in the absence of oxygen), or facultative (can live on oxygen and live without it through different metabolisms).

Aerobic biological treatment dominates secondary wastewater treatment scenes, and it is performed in the presence of oxygen by aerobic microorganisms (principally bacteria) that metabolize the organic matter in the wastewater, thereby producing more microorganisms and inorganic end products (principally carbon dioxide, ammonia and water). Several aerobic biological processes are used for secondary treatment, differing primarily in the manner in which oxygen is supplied to the microorganisms and in the rate at which organisms metabolize the organic matter. From a nutritional point of view, the majority of microorganisms in biological wastewater treatment systems use the organic matter in the wastewater as their energy source for growth and maintenance.

Sometimes, anaerobic processes are also used in the secondary biological treatment of wastewater. Anaerobic processes, in addition to sludge digestion, are employed to treat high-strength wastewater, such as high-strength food

Food and Agricultural Wastewater Utilization and Treatment, Second Edition. Sean X. Liu.
© 2014 John Wiley & Sons, Ltd. Published 2014 by John Wiley & Sons, Ltd.

processing wastewater streams, when the prospect of difficulty associated with oxygen supply to the reactor and large biomass produced in an aerobic process is deemed uneconomical.

In secondary wastewater treatment, sedimentation is also employed to remove settleable solids after the microorganisms have done their work. The microorganisms must be separated from the treated wastewater by sedimentation to produce clarified secondary effluent. The sedimentation tanks used in secondary treatment, often referred to as secondary clarifiers, operate in the same basic manner as the primary clarifiers described in Chapter 3. The biological solids removed during secondary sedimentation – called secondary or biological sludge – are normally combined with primary sludge for subsequent sludge processing.

The main difference between sedimentation in a secondary treatment reactor (tank/basin) and sedimentation in a primary treatment reactor (tank/basin) is that sludge in the secondary treatment is comprised of biological cells. There are two main types of bioreactors used in the secondary treatment: those where microorganisms are attached to a fixed surface (e.g., trickling filter), and those where microorganisms run freely in the wastewater stream (e.g., activated sludge). The sludge that has settled in the sedimentation tank in the latter type of reactors is usually recycled back into the system for continuing operations.

Widespread high-rate processes in relatively small reactors include the activated sludge processes, trickling filters, and rotating biological contactors. A combination of two of these processes in series (e.g., trickling filters followed by activated sludge) is sometimes used for certain wastewaters containing a high concentration of organic materials from food and agricultural processing.

Since a secondary treatment process uses microorganisms to break down the organic matters in order to clarify the wastewater, it is important to know the biology of the process. The most widely present microorganisms in wastewater treatment are bacteria, and this group of microorganisms is responsible for degrading organic matter present in the wastewater. The following section, about kinetics of biochemical systems in wastewater treatment systems, is centered on the role of bacteria in biological wastewater treatment. However, the essence of this is applicable to all microorganisms in biological conversions in biochemical systems.

4.2 Kinetics of biochemical systems in wastewater microbiology

The fundamentals of wastewater microbiology include the roles of microbial groups in specific biological transformation of wastewaters, nutritional

4.2 KINETICS OF BIOCHEMICAL SYSTEMS IN WASTEWATER MICROBIOLOGY 105

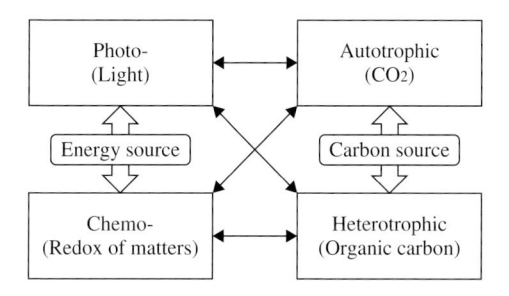

```
┌──────────────┐                    ┌──────────────┐
│   Photo-     │◄──────────────────►│  Autotrophic │
│   (Light)    │                    │    (CO2)     │
└──────────────┘        ╲╱          └──────────────┘
       ▲         ╲                         ▲
  ┌─────────┐     ╲        ┌──────────┐
  │ Energy  │      ╲╱      │  Carbon  │
  │ source  │      ╱╲      │  source  │
  └─────────┘     ╱         └──────────┘
       ▼         ╱                         ▼
┌──────────────┐        ╱╲         ┌──────────────┐
│   Chemo-     │◄──────────────────►│ Heterotrophic│
│(Redox of matters)│                │(Organic carbon)│
└──────────────┘                    └──────────────┘
```

Figure 4.1 Classification of microorganisms based on nutritional requirements

requirements, the effects of environmental conditions on microbial activities, and enzymatic reactions that underpin biological conversions of waste materials.

We have discussed the classifications of various microbial groups and their roles in biological wastewater treatment, based on cell structures and their function as eukaryotes and prokaryotes, in Chapter 2. However, microorganisms in wastewater treatment can also be described on the basis of their nutritional requirements. Like all living things, nutrients play a critical role in development of microbial communities. They supply the energy source for cell growth and biosynthesis, and provide the materials necessary for the synthesis of cytoplasmic materials, as well as serving as electron acceptors from biochemical reactions. The nutritional requirements provide a basis of microorganism classifications based on carbon source and energy source. Figure 4.1 illustrates a general classification of microorganisms based on these nutritional requirements.

Environmental factors, such as temperature, pH, and oxygen requirements for aerobic or facultative microorganisms, are of great importance to microbial growth or even survival. The temperature effect on microbial growth tends to be positive, although overheating may inhibit or kill microorganisms. The effects of pH on microbial communities are more varied; some microorganisms do well in slightly alkaline conditions such as most bacteria with pH ranging from 6.5–7.4, but fungi prefer slightly acidic conditions. Oxygen requirements are critically important for aerobes and are optional for facultative bacteria, while anaerobes are not adversely affected at all by the absence of oxygen; they thrive without it.

Bioconversion of organic matter by microorganisms takes place in a series of biochemical reactions with the participation of enzymes. Enzymes are specific proteins that catalyze reactions but do not undergo permanent changes themselves. They work by forming complexes with organic substrate and inorganic molecules, and facilitating the reaction of these substances, resulting in end products and releasing the enzymes in their original state so that the biochemical reaction cycle can continue.

Enzymes are substrate-specific and, thus, bacteria usually have many different enzymes performing different catalytic roles in converting a broth of organic substances into end products. In general, these enzymes belong to one of two groups of enzymes: extracellular and intracellular. As the names suggest, extracellular enzymes convert organic substances outside the cell into a form of intermediate products; intracellular enzymes can then take over from there to complete the biochemical reactions within the cells. These enzymatic reactions often occur sequentially among different enzymes in a cell. A portion of the organic substrate that attaches to the enzyme is utilized as an energy source, while the remainder is scavenged to reproduce more cells.

The microbial population of biological wastewater treatment systems contains a large number of species of microorganisms with diverse physiological and genetic variations. As a result, the properties of the colony in wastewater systems may only be described as averaged behaviors of the microbial population. Additionally, the properties of the microbial population are described in terms of easily quantifiable parameters. For example, the size of the population is often measured in dry weight or nitrogen content, since microbial colonies are established along the line of functionally discrete units or cell mass. Thus, the microbial population in wastewater treatment systems is considered as a mixture of microorganisms as biomass distributed continuously in treatment systems (or reactors). With that view, we can treat biochemical reactions involving microbial population in reactors as averaged reactants (biomass and organic substrate) undergoing enzymatic reactions. The models that describe characteristics of these abstract "reactants" are therefore *deterministic*, despite the fact that the behaviors of individual cells within the microbial community can only be described as *stochastic*.

The kinetic models that describe the biological conversion of organic matter in wastewater represent the reactions that result in changes in concentration of an organic substrate or microorganism responsible for the conversion, and they may be modeled using simple reaction rate theory, which describes the rate of change in concentration of an organic substrate or microorganism:

$$A \rightarrow B$$
$$\text{Reactant} \quad\quad \text{Product} \tag{4.1}$$

$$\text{rate} = k\,(\text{concentration of }A)^n \tag{4.2}$$

The rate of reaction can further be expressed as:

$$-r = kC_A^n \tag{4.3}$$

4.2 KINETICS OF BIOCHEMICAL SYSTEMS IN WASTEWATER MICROBIOLOGY

where:
> k is the reaction rate constant
> A is concentration of species A (substrate or biomass)
> n is the order of the reaction.

The negative sign in Equation (4.3) signifies the disappearing of reactant A as the reaction progresses. The order of reaction is a parameter that reflects the kinetics of a biochemical reaction and can theoretically be any number, but often it is one of the three basic reaction kinetics – namely, zero-order, first-order, and second-order, as described in Chapter 1.

Taking log on both sides of Equation (4.2) yields

$$\log r = n \log(\text{concentration of reactant A}) + \log k \quad (4.4)$$

A plot of $\log r$ vs. $\log C_A$ will produce a linear line with a slope of n.

4.2.1 Effects of temperature on reaction rates

Like any other reactions, the effects of temperature on reaction rates originate from the temperature effects on rate constants. The rate constant, k, is a lumped parameter that encompasses many environmental factors such as pH, oxygen concentration, concentrations of trace elements, and in the case of photosynthesis, light intensity. One particularly important parameter is temperature, which affects the rates of both chemical and biochemical reactions. It was observed by Van't Hoff that a reaction rate roughly doubles for every 10°C increase in temperature. The effects of temperature on reaction rates often follow the Arrhenius model:

$$k = Ae^{-\frac{E_A}{RT}} \quad (4.5)$$

where:
> T is absolute temperature or thermodynamic temperature in Kelvin
> E_A is activated energy that reactants must overcome in order to proceed with reactions
> R is the gas constant ($R = 8.314 \text{ J K}^{-1} \text{ mol}^{-1}$)
> A is pre-exponential factor or frequency factor that is related to collision frequency of reactants.

The Arrhenius equation is widely used in wastewater treatment systems to model the effects of temperature on reactions. It first starts with taking a derivative of

Equation (4.5) and then integrating between the limits T_0 and T_f; this gives

$$\frac{k_f}{k_0} = \left(\frac{E_A}{RT_f T_0}\right)^{(T_f - T_0)} \quad (4.6)$$

where k_0 and k_f are the rate constants at temperatures of T_0 and T_f, respectively.

The reason for taking a derivative and following with integration is that the rate constant is a function of time. Equation (4.6) has been used in the temperature range of 5–25°C but, outside this range, the equation is not valid, due to significant changes in composition in the microbial population.

4.2.2 Effects of pH and dissolved oxygen concentration on reaction rates

Effects of pH and dissolved oxygen concentration on reaction rates are more substrate-specific. The optimum pH range for carbonaceous oxidation lies in the limits of 6.5–8.5. At a pH above 9.0, microbial activity is inhibited; at a pH below 6.5, fungi dominate over the bacteria in the competition for the substrate. Some types of reactors are less affected by fluctuations of influent pH and completely mixed reactors, such as CSTRs, will minimize the effect of pH fluctuation. If the pH fluctuation is significant, some adjustment to pH may be needed. Some bacteria have less tolerance towards pH fluctuation than others; for example, anaerobic bacteria have a viable pH range of 6.7–7.4, with optimum growth occurring in pH 7.0–7.1.

Dissolved oxygen (DO) concentration obviously affects rates of reactions in aerobic biochemical reactions. For instance, a DO concentration of 1–2 mg/L may be sufficient for active aerobic heterotrophic microbial activity, provided that sufficient nutrients and trace elements are available to microbial activities.

4.2.3 Kinetic equations of bacterial growth

The kinetics of biochemical reactions in bioreactors described in the previous sections, and in Chapter 1, are simplified mathematical descriptions that are not all-inclusive in terms of all aspects of the mechanisms under consideration. Successful environmental control in biological wastewater treatment, however, is rooted in an understanding of the basic principles governing the growth of microorganisms where substrates are assimilated and biomass in the system accumulates. The studies of microorganism growth in pure media under controlled pH, temperature, oxygen concentration, and other substances, have produced reliable kinetic models of microorganism growth.

4.2 KINETICS OF BIOCHEMICAL SYSTEMS IN WASTEWATER MICROBIOLOGY

The general growth pattern of bacteria in a batch culture is illustrated in Figure 4.2. This diagram is a record of number of viable microorganisms in a culture medium of fixed volume over time. The pattern of the curve shown in Figure 4.2 shows four distinct phases: the lag phase, the log-growth phase, the stationary phase, and the log-death phase.

- *The lag phase.* Once the inoculum is introduced in the culture medium, the microorganisms take time to acclimatize themselves in the environment.

- *The log-growth phase.* This is the normal growth pattern under sufficient food (organic substrate) and nutrients for microbial growth

- *The stationary phase.* The microbial population is stabilized as a result of a standoff between growth of microorganisms and death of old cells. Usually, there is insufficient substrate or/and nutrients available for microbial growth.

- *The log-death phase.* During this phase, the death rate of microbial cells exceeds the growth rate of new cells. This is an indication of deteriorating environmental conditions in addition to a lack of substrates.

The common autocatalytic equation, a first-order reaction, is used to describe the log-growth phase:

$$r_X = \frac{dX}{dt} = \mu X \tag{4.7}$$

where:
r_X is the rate of production of viable bacteria
X is the concentration of viable microorganisms
μ is the specific growth rate constant (t^{-1}).

If the cell concentration of microbial cells at t_0 is X_0, and after a time interval, t, the viable cell concentration X_t is:

$$X_t = X_0 e^{\mu t} \tag{4.8}$$

Taking logs of Equation (4.8) shows a linear relationship between $\ln X$ and $\ln X_0$.
When cell concentration doubles between t_d and t_0, it shows the following relationship:

$$2X_0 = X_0 \exp(\mu t_d) \quad 2 = \exp(\mu t_d) \quad \ln 2 = \mu t_d \tag{4.9}$$

There are several models that relate cell growth to substrate utilization, and the Monod model is the most widely used model. The Monod model describes the relationship between the residual concentration of the cell-growth limiting

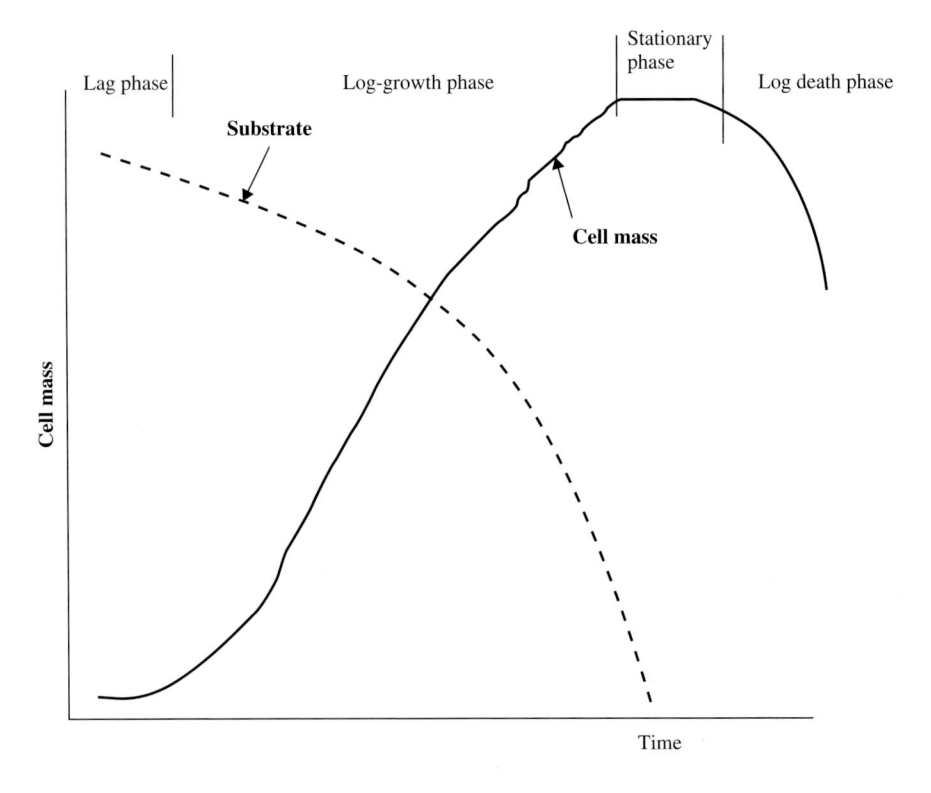

Figure 4.2 Diagram of number of viable microorganisms in the culture medium of fixed volume over time

substrate or nutrient and the specific growth rate of biomass of cells, μ, in the following mathematical expression:

$$\mu = \mu_m \frac{S}{S + k_s} \quad (4.10)$$

where:
μ_m is the maximum specific growth rate at saturation concentration of the growth-limiting substrate
S is the substrate concentration
k_s is the saturation constant (mg/L), which is the concentration of growth-limiting substrate at which the specific growth rate $\mu = \mu_m/2$.

4.3 Idealized biochemical reactors

Almost all bioreactors used for kinetics studies, whether batch or continuous reactors, are types of well-mixed reactors. Two well-mixed reactors are

4.3 IDEALIZED BIOCHEMICAL REACTORS

briefly described below: the ideal batch reactor and the ideal continuous-flow stirred-tank reactor (CSTR).

4.3.1 The ideal batch reactor

In a batch reactor, the concentrations of nutrients, substrates, products, and bacterial cells vary with time as the microbial growth proceeds.

A molar material balance on component A in a batch reactor yields the following relationship:

$$\frac{d}{dt}\left[\left(\begin{array}{c}\text{culture}\\ \text{volume}\end{array}\right)\left(\begin{array}{c}\text{molar concentration}\\ \text{of component A}\end{array}\right)\right]$$
$$= \left(\begin{array}{c}\text{culture}\\ \text{volume}\end{array}\right) \cdot \left(\frac{\text{moles A formed by reaction}}{\text{unit culture volume} \cdot \text{unit time}}\right) \quad (4.11)$$

Or, mathematically:

$$\frac{d}{dt}(V_R \cdot C_A) = V_R \cdot r_A \quad (4.12)$$

where:
V_R is the culture volume
C_A is the molar concentration
r_A is the reaction rate.

The volume of the batch reactor is not the culture volume, V_R, unless the reactor is full. V_R is often considered as constant so long as there is no addition or removal of liquids from the reactor. This fact leads to the following simplification:

$$\frac{dC_A}{dt} = r_A \quad (4.13)$$

In order to calculate the change in concentration of component A, Equation (4.13) needs to be integrated over the time period, $t-t_0$:

$$\int_{C_{A0}}^{C_A} \frac{dC_A}{r_A} = \int_{t_0}^{t} dt = t - t_0 \quad (4.14)$$

where C_{A0} is the concentration of A at time t_0.

The function that relates r_A to t can be obtained based on the order of the reaction, and the kinetic equations of biochemical reactions described in Chapter 1 can be used to solve Equation (4.14).

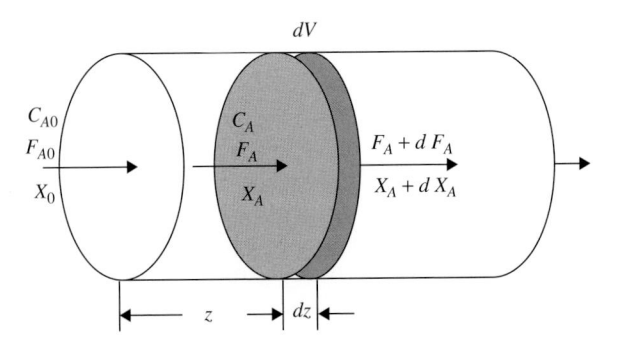

Figure 4.3 A schematic diagram of a plug-flow reactor

4.3.2 Ideal plug-flow reactor

The plug-flow reactor is also called a piston-flow reactor. All materials leaving the plug-flow reactor will have been in the reactor for the same length of time and it is assumed that there is no longitudinal mixing or diffusion. A mathematical derivation of a plug-flow reactor, as shown in Figure 4.3, can be made based on material balance for an element of reactor volume, dV, for a component A:

input − output = disappearance due to reaction
input = F_A
output = $F_A + dF_A$
disappearance = $-r_A dV$

$$F_A = F_A + dF_A + (-r_A)dV \tag{4.15}$$

where F_A is the flow rate of material A.
Since:

$$dF_A = d[F_{A0}(1-X_A)] = -F_{A0}dX_A : \tag{4.16}$$

Equation (4.15) becomes:

$$F_{A0}dX_A = (-r_A)dV \tag{4.17}$$

$$\frac{dV}{F_{A0}} = \frac{dX_A}{-r_A} \tag{4.18}$$

This can be integrated to produce the governing equation for a plug-flow reactor:

$$\frac{C_{A0}V}{F_{A0}} = r_A = C_{A0}\int_0^{X_A} \frac{dX_A}{-r_A} \tag{4.19}$$

where $C_{A0} = F_{A0}/V$ and r_A can be a zero, first, or second order reaction.

4.3 IDEALIZED BIOCHEMICAL REACTORS

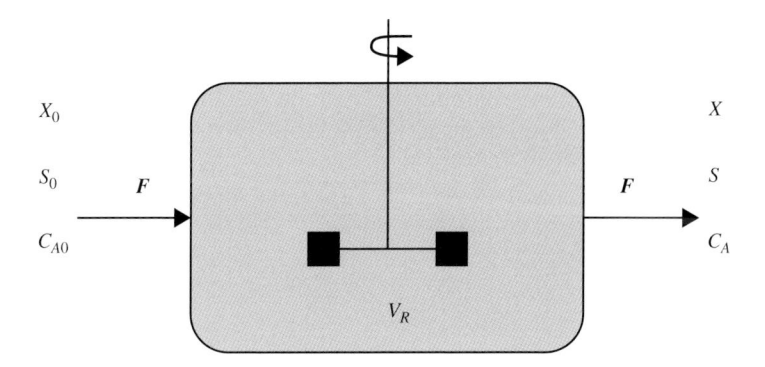

Figure 4.4 Schematic diagram of a CSTR type of reactors

4.3.3 Ideal continuous-flow stirred-tank reactor (CSTR)

Figure 4.4 schematically depicts a CSTR type of reactors. In CSTRs, the liquid inside the reactor is completely mixed. The mixing is provided through an impeller, rising gas bubbles (usually oxygen) or both. The most characteristic feature of a CSTR is that it is assumed that the mixing is thorough and complete such that the concentrations in any phase do not change with position within the reactor.

As indicated in Figure 4.4, the dissolved oxygen in the tank is the same throughout the bulk liquid phase. Because of this uniformity of oxygen distribution in the reactor, a CSTR for wastewater treatment operations has the advantage of de-coupling the aerator or stirrer from the reaction so long as oxygen is well provided for (no need to consider pesky fluid mechanics), thus simplifying process design and optimization. Under the steady state, where all concentrations within the reactor are independent of time, we can apply the following materials balance on the reactor:

$$\begin{bmatrix} \text{Rate of addition} \\ \text{to reactor} \end{bmatrix} + \begin{bmatrix} \text{Rate of accumulation} \\ \text{within reactor} \end{bmatrix} = \begin{bmatrix} \text{Rate of removal} \\ \text{from reactor} \end{bmatrix} \quad (4.20)$$

Replacing the statements in the above expression with mathematical symbols leads to:

$$FC_{A0} + V_R r_A = FC_A \qquad F(C_{A0} - C_A) = -V_R r_A \quad (4.21)$$

where F is volumetric flow rate of feed and effluent liquid streams.

Rearrangement of Equation (4.21) yields:

$$r_A = \frac{F}{V_R}(C_A - C_{A0}) = D(C_A - C_{A0}) \quad (4.22)$$

where $D = F/V_R$ and is called "dilution rate". The term characterizes the holding time or processing rate of the reactor under steady state conditions. It is the number of tank-full volumes passing through the reactor tank per unit time and is equal to the reciprocal of the mean holding time of the reactor.

Because of lack of time dependence of concentrations in CSTR and, thus, the differential form of reactor analysis as in a batch reactor, CSTRs have the advantage of being well-defined, easily reproducible reactors that are used frequently in many cell growth kinetics studies, despite relatively high cost and a long time for achieving steady state. Batch reactors, which can be as simple as a sealed beaker or flask in an incubator shaker, are still widely used for their inexpensive, quick, and unbridled benefits. However, no matter what type of reactors are used, the goal of studying cell growth kinetics should be based on the intended application and scope of the use of the kinetics. Only then may the experimental design and implementation be formulated.

4.4 Completely mixed aerated lagoon (CMAL)

Lagoons are one of the oldest wastewater treatment systems created by humankind. They consist of lined in-ground earthen basins, in which the wastewater is detained for a specified time (detention time) and then discharged. The size and depths can vary, as well as the degree of treatment. Although these lagoons – or ponds, as they are sometimes called – are very simple in design, there are complex chemical, biological and physical processes occurring. There are five main types of lagoons: facultative lagoon (stabilization pond); aerobic lagoon; anaerobic lagoon; partial mixed aerated lagoon; and completely mixed aerated lagoon.

A completely mixed aerated lagoon in wastewater treatment is a relatively shallow basin (with a depth between 5–15 feet, or between 13–38 cm) and a large surface area (of several acres), which operates on a flow-through mode, similar in many ways to CSTRs. They are designed and operated to exclude algae by completely mixing the solids and, therefore, blocking all light. The earthen basin is generally lined with impervious material such as asphalt or plastic. Complete mixing is achieved through the use of mechanical surface aerators of either the fixed or floated type. The mixing energy must be sufficient so that all solids are suspended and will reduce light in the wastewater to the extent that very little algal growth occurs, since control of algal growth is crucial in the reduction of effluent suspended solids. The holding time in a CMAL is typically 7–10 days. The degree of treatment is a function of the mass of organic matter in suspension and holding time.

4.4 COMPLETELY MIXED AERATED LAGOON (CMAL)

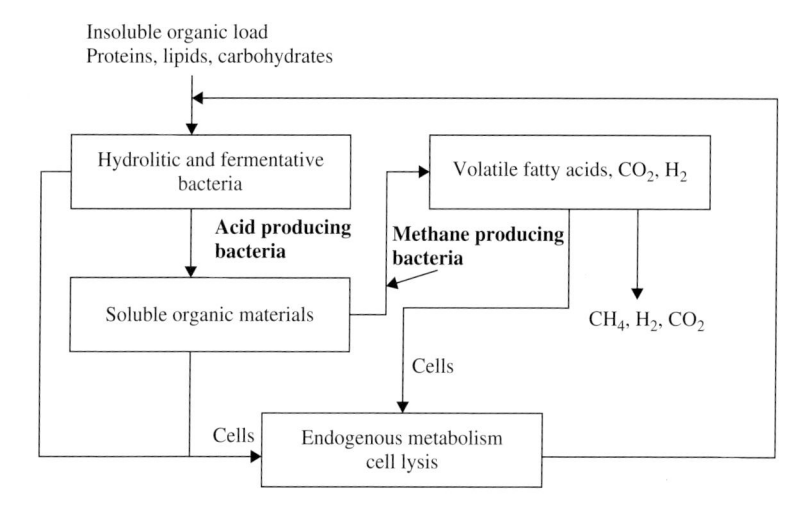

Figure 4.5 Photograph of a typical CMAL installation

The microbial population in a CMAL is predominately heterotrophic, and aerobic respiration is the path of metabolism of the microbial population in excess of oxygen. The bacteria in a CMAL utilize the organic matter, and life forms higher in the food chain, such as protozoans, rotifers, daphnia, and insect larvae, are likely to feed on these bacteria and their predators. The ecology of a CMAL is rather complex and may also involve growth of algae, although algae growth should be suppressed. This issue, and other possible interactions within the flora of a CMAL, should be an important factor during the treatability studies of wastewater streams. A photograph of a typical CMAL installation is shown in Figure 4.5.

In a typical CMAL treatment system, a series of lagoons are placed to treat the expected ranges of food and agricultural wastewater. The number of lagoons can vary, as can the sizes. When CMALs are used for wastewater treatment before land applications, the first successive lagoons provide biological treatment of the wastewater and the last lagoon provides storage. CAMLs are usually employed to achieve one of two objectives: degrading soluble solids into insoluble biomass (microbial cells); and stabilizing the organic solids. CMALs are used as a pre-treatment for industrial wastewaters, or as total wastewater treatment systems for a small community.

The main advantage to using lagoon systems such as CMALs is their simplicity of treatment. Lagoons treat wastewater over an extended period of time and can be designed to degrade sludges completely. They are not as susceptible to "shock" loading; for example, should a toxic chemical or high pH load be introduced into the lagoon system, the constituent is diluted and can be isolated.

The regular wastewater can be bypassed to the next lagoon for continued treatment. Lagoon systems also help to equalize peak inflows.

Lagoon systems are simple, low-capital investment systems and also have low operation and maintenance costs, since fewer staff and less mechanical equipment are needed to operate this type of system. Minimal sludge is produced, which reduces capital and operational costs associated with sludge stabilization, conditioning, dewatering and disposal. The main disadvantages of lagoon systems are that they require more land than mechanical wastewater treatment facilities and are thus not a viable option for communities with high population intensity. Also, there is no operational control over the rates of biochemical reactions; the degree of treatment is affected by the temperature, as temperature effects the kinetics of biochemical reactions. Seasonal changes tend to have impact on the degree of the treatment with CMALs.

Completely mixed aerated lagoons offer a reasonable treatment alternative to more costly mechanical biological wastewater treatment such as trickling filters and activated sludge systems for readily biodegradable wastewater – those streams from food processing operations. However, the decision whether to use CMALs or not should be based on the total cost analysis, treatability study, and local environmental regulations.

4.5 Trickling filter (TF)

Trickling filters (also called biofilters) have been used to remove organic matter from wastewater for nearly 100 years. The TF is an aerobic treatment system that utilizes microorganisms attached to a medium to remove organic matter from wastewater. The colonies of microorganisms attached to solid surfaces are called biofilms, due to their thin layers of biological structures. This type of system is common to a number of biological wastewater technologies, such as rotating contactors and packed bed reactors (also called bio-towers).

TFs are mainly comprised of four major components: a filter medium such as stones, plastic shapes, or wooden slats; an enclosure to hold the liquid; a distribution system; and an under-drain system. The filter medium provides the surface upon which the microorganisms grow. The enclosure holds both wastewater and filter medium, while the distribution system ensures a uniform hydraulic load over the entire TF and the under-drain system provides drainage, holds the filter medium and supplies oxygen to the bottom section.

4.5.1 How TFs work

A rotary or stationary distribution system distributes wastewater from the top of the filter, percolating it through the interstices of the medium (see Figure 4.6). As

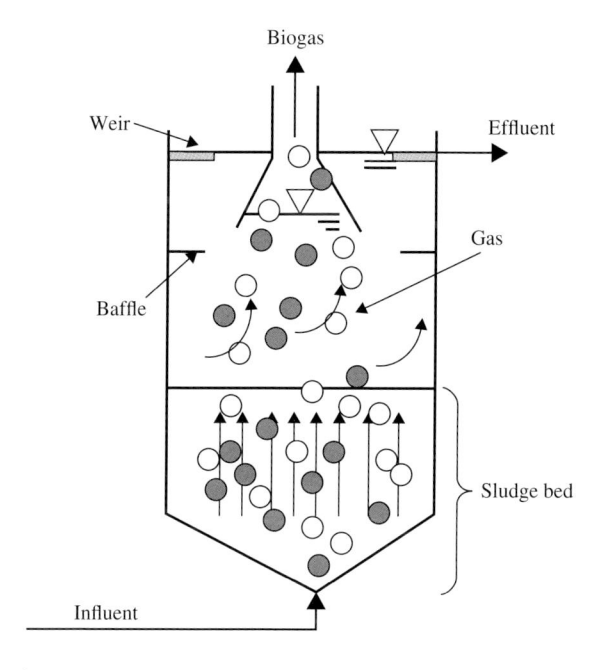

Figure 4.6 Schematic diagram of a trickling filter system

the wastewater flows over the medium, the organic matter in the wastewater is adsorbed by a population of microorganisms (aerobic, anaerobic, and facultative bacteria; fungi; algae; and protozoa) attached to the medium as a biological film or slime layer (approximately 0.1 to 0.2 mm thick).

As the wastewater flows over the medium, microorganisms already in the water gradually attach themselves to the rock, crushed granite, or plastic structure surface and form a film. The aerobic microorganisms in the outer part of the slime layer (biofilm) then decompose the organic material. As the layer thickens through microbial growth, oxygen cannot penetrate the medium surface, and anaerobic organisms prosper. As the biological film continues to grow, the microorganisms near the surface lose their ability to cling to the medium, and a portion of the slime layer falls off the filter. This process is known as sloughing. The sloughed solids (sludge) are collected by the under-drain system and transported to a clarifier for removal from the wastewater.

4.5.2 Advantages and disadvantages

There are advantages and disadvantages of TFs associated with biological treatment of food and agricultural wastewater. Their importance depends on the needs of the end user and the characteristics of the wastewater.

118 CH04 BIOLOGICAL WASTEWATER TREATMENT PROCESSES

Advantages include:

- Simple, reliable, biological process.
- Suitable in areas where large tracts of land are not available for land intensive treatment systems.
- May qualify for equivalent secondary discharge standards.
- Effective in treating high concentrations of organic matters depending on the type of medium used.
- Rapidly reduces soluble BOD_5 in wastewater streams.
- Efficient nitrification units.
- Low power requirements.
- Moderate level of skill and technical expertise needed to manage and operate the system.

Disadvantages include:

- Additional treatment may be needed to meet more stringent discharge standards.
- Possible accumulation of excess biomass that cannot retain an aerobic condition and can impair TF performance.
- Requires regular operator attention.
- Incidence of clogging is relatively high.
- Requires low loadings depending on the medium.
- Flexibility and control are limited in comparison with activated-sludge processes.
- Odor problems.
- Snail problems.

4.5.3 Design criteria

A TF consists of permeable medium made of a bed of rock, slag, or plastic over which wastewater is distributed to trickle through, as shown in Figure 4.6. Depending on the properties of wastewater streams and treatment requirements (see Table 4.1 for TF design parameters), rock or slag beds can be up to 60.96 m (200 ft) in diameter and 0.9–2.4 m (3–8 ft) deep, with rock size varying from 2.5–10.2 cm (1–4 in). Most rock media provide approximately 149 m^2/m^3 (15 sq ft/cu ft) of surface area and less than 40% void space. Packed plastic filters (bio-towers), on the other hand, are smaller in diameter (6–12 m (20–40 ft)) and range in depth from 4.3–12.2 m (14–40 ft). These filters look more like towers, with the media in various configurations (e.g., vertical flow, cross flow, or various random packings).

Research has shown that cross-flow media may offer better flow distribution than other media, especially at low organic loads. When comparing vertical media with the 60-degree cross-flow media, the vertical media provide a nearly equal distribution of wastewater, minimizing potential plugging at higher organic loads better than cross flow media. The plastic medium also requires additional provisions, including ultraviolet protective additives on the top layer of the plastic medium filter, and an increased plastic wall thickness for medium packs that are installed in the lower section of the filter, where loads increase.

The design of a TF system for wastewater also includes a distribution system. Rotary hydraulic distribution is usually standard for this process, but fixed nozzle distributors are also being used in square or rectangular reactors. Overall, fixed nozzle distributors are being limited to small facilities and package plants. Recently, some distributors have been equipped with motorized units to control their speed. Distributors can be set up to be mechanically driven at all times or during stalled conditions.

In addition, a TF has an under-drain system that collects the filtrate and solids and also serves as a source of air for the microorganisms on the filter. The treated wastewater and solids are piped to a clarifier and the sludge is removed from the bottom of the clarifier for further treatment or disposal. The clarified treated water is either discharged or recycled back to mix with the influent of the trickling filter system.

4.6 Rotating biological contactor (RBC)

Rotating Biological Contactors are used in the treatment of wastewater as a secondary treatment process. The RBC process involves allowing wastewater to come in contact with a biological medium in order to remove contaminants in sewage before discharging the treated wastewater to the environment, usually a river.

The construction of a rotating biological contactor consists of a series of plastic discs (the media) mounted on a driven shaft that is contained in a tank or trough. Commonly used plastics for the media are polythene, PVC, and expanded polystyrene. The shaft is aligned with the flow of sewage, such that the discs rotate at right angles to the flow, with several rotors usually combined to make up a treatment train. About 40% of the disc area is immersed in the sewage.

Table 4.1 Trickling filter types in biological wastewater treatment

Filter type	Filter medium	Hydraulic loading (gal/ft²–min)	BOD₅ loading (lb/ft³–day)	BOD5 removal (%)	Depth (ft)	Re-circulation ratio	Film sloughing	Nitrification
Low rate	Rock & slag	0.0–0.06	0.005–0.025	80–90	6–8	0	Intermittent	Well
Intermediate rate	Rock & slag	0.06–0.16	0.015–0.03	50–70	6–8	0–1	Intermittent	Partial
High rate	Rock	0.16–0.64	0.03–0.06	65–85	3–6	1–2	Continuous	Little
Super high rate	Plastic	0.2–1.2	0.03–0.1	65–80	10–40	1–2	Continuous	Little
Roughing	Plastic & redwood	0.8–3.2	0.1–0.5	40–65	15–40	1–4	Continuous	None
Two–stage	Rock & plastic	0.16–0.64	0.06–0.12	85–95	6–8	0.5–2	Continuous	Well

Source: Nguyen & Shieh (2000).

4.6 ROTATING BIOLOGICAL CONTACTOR (RBC)

The biological growth that becomes attached to the discs assimilates the organic materials in the wastewater. Aeration is provided by the rotating action, which exposes the media to the air after contacting them with the wastewater, facilitating the digestion of the organic compounds that need to be removed. The degree of wastewater treatment is related to the amount of media surface area and the quality and volume of the inflowing wastewater.

The RBC process may be used where the wastewater is suitable for biological treatment and can be used in many modes to accomplish varied degrees of carbonaceous and/or nitrogenous oxygen demand reductions. The process is simpler to operate than activated sludge, since recycling of effluent or sludge is not required. Special consideration must be given to returning supernatant from the sludge digestion process to the RBCs. The advantages of RBC technology include a longer contact time (8–10 times longer than trickling filters), a higher level of treatment than conventional high-rate trickling filters, and less susceptibility to upset from changes in hydraulic or organic loading than the conventional activated sludge process.

Whether used in small or large facilities, the RBC process should be designed to remove at least 85% of the BOD from domestic sewage. The process can also be designed to remove ammonia nitrogen (NH_3-N). In addition, the RBC process can treat effluents and process wastewater from dairies, bakeries, food processors, pulp and paper mills, and other biodegradable industrial discharges.

4.6.1 Process selection

Choice of the process mode most applicable will be influenced by the degree and consistency of treatment required, the type of waste to be treated, site constraints, and capital and operating costs. The process design of a RBC facility involves an accurate determination of influent and side-stream loadings, proper media sizing, staging and equipment selection to meet effluent requirements, air requirements, and selection of an overall plant layout that will provide for flexibility in operation and maintenance.

A comprehensive on-site pilot plant evaluation is recommended to incorporate the factors affecting RBC performance as an accurate source of information for a RBC design. Other approaches to determine the expected performance of RBCs may be based upon results of similar full-scale installations and/or through documented pilot testing with the particular wastewater. Small-diameter RBC pilot units are suitable for determining the treatability of the wastewater. If small-diameter units are operated to obtain design data, each stage must be loaded below the oxygen transfer capability of a full-scale unit to minimize scale-up problems. Direct scale-up from small-diameter units to full-scale units is not possible because of the effects of temperature, peripheral speed of media, and other process and equipment factors.

In all RBC systems, the major factors controlling treatment performance are:

- organic and hydraulic loading rates;
- influent wastewater characteristics;
- wastewater temperature;
- biofilm control;
- dissolved oxygen levels; and
- flexibility in operation.

4.6.2 Pretreatment

Raw wastewater should not be applied to an RBC system; primary settling tanks are required for effective removal of grit, debris, and excessive oil or grease prior to the RBC process. In some cases, fine screens (0.076–0.152 cm or 0.03–0.06 inches) may be considered. Screening and comminution are not suitable as the sole means of preliminary treatment ahead of RBC units.

Sulfide production must be considered in the system design. Separate facilities to accept and control feeding of in-plant side streams should be considered where the potential for sulfide production or increased organic and ammonia nitrogen loadings will have a significant impact on the RBC system.

4.6.3 Design criteria

Unit sizing

Organic loading is the primary design parameter for the RBC process. This is generally expressed as the organic loading per unit of media surface area per unit of time, or in units of pounds BOD_5 per thousand square feet per day. Wastewater temperatures above 13°C or 55°F have minimal affect on organic removal and nitrification rates but, below 13°C or 55°F, manufacturers need to be contacted to obtain the various correction factors that must be utilized to determine the needed additional media surface area. In determining design-loading rates on RBCs, the following parameters should be utilized.

4.6 ROTATING BIOLOGICAL CONTACTOR (RBC)

Design flow rates and primary wastewater constituents:

- total influent BOD_5 concentration;
- soluble influent BOD_5 concentration;
- percentage of total and soluble BOD_5 to be removed;
- wastewater temperature;
- primary effluent dissolved oxygen;
- media arrangement, number of stages, and surface area of media in each stage;
- rotational velocity of the media;
- retention time within the RBC tank(s);
- influent soluble BOD_5 to the RBC system including soluble BOD from in-plant side-streams, etc.;
- influent hydrogen sulfide concentrations; and
- peak loading, BOD_5 max/BOD_5 avg.

In addition to the above parameters, loading rates for nitrification will depend upon influent DO concentration, influent ammonia nitrogen concentration and total Kjeldahl nitrogen (TKN), diurnal load variations, pH and alkalinity, and the allowable effluent ammonia nitrogen concentration.

Since soluble BOD_5 loading is a critical parameter in the design of RBC units, it should be verified by influent sampling whenever possible.

Loading rates

When peak to average flow ratio is 2.5 to 1.0 or less, average conditions can be considered for design purposes. For higher flow ratios, flow equalization should be considered.

The organic loading to the first stage standard density media should be in the range of 3.5–6.0 pounds total BOD_5 per thousand square feet per day (1 pound BOD per thousand square feet per day is about 4.9 gram BOD per square meter per day), or 1.5–2.5 pounds soluble BOD_5 per thousand square feet per day. First stage organic loadings above 6 pounds total BOD_5 or 2.5 pounds soluble BOD_5 per thousand square feet per day will increase the probability of developing problems such as excessive biofilm thickness, depletion of dissolved oxygen, nuisance organisms and deterioration of process performance. The most critical problem in most instances is the structural overloading of the RBC shaft(s).

For average conditions, the design loading should not exceed 2.5 pounds of soluble BOD_5/1,000 square feet of standard media surface per day on the first stage shaft(s) of any treatment train. Periodic high organic loadings may require supplemental aeration in the first stage shafts. High density media should not be used for the first stage RBCs.

For peak conditions, the design loading shall not exceed 2.0 pounds of soluble BOD_5/1,000 square feet for the first high density media shaft(s) encountered after the first two shafts or rows of shafts in a treatment train.

For average conditions, the overall system loading should not exceed 0.6 pounds of soluble BOD_5/1,000 square feet of media. This soluble BOD_5 loading to all shafts should be used to determine the total number of shafts required. Equation (4.23) in the later section could be used as an option to determine the number of stages required.

Staging units

Staging of RBC media is recommended to maximize removal of BOD and ammonia nitrogen (NH_3-N). In secondary treatment applications, rotating biological contactors need to be designed with a minimum of three stages per flow path. For combined BOD_5 and NH_3-N removal, a minimum of four stages is recommended per flow path. For small installations, multiple stages are acceptable on a single shaft if inter-stage baffles are installed within the tank and the flow is introduced parallel to the shaft. Whenever multiple process trains are employed with three or more shafts in a row, the flow path should be introduced perpendicular to the shafts and the wastewater should be distributed evenly across the face of the RBCs.

The organic loading must be accurately defined by influent sampling whenever possible. For existing facilities that are to be expanded and/or rehabilitated, it is unacceptable only to calculate the expected load to the shafts. Flow and load sampling must be done to demonstrate the load that is generally accomplished by composite sampling after primary clarification. To predict effluent quality for a range of loadings, the influent and effluent soluble-to-total BOD_5 ratio can be assumed to be 0.5.

An alternative method of estimating soluble organic removal in the inter-stages, devised by Opatken (1986), utilizes a second order reaction equation. The equation may be used for RBC design during the summer months, but a temperature correction factor should be used for the cold winter months. Wastewater temperatures below 15°C decrease shaft rotational speeds and increase loping problems, resulting in insufficient biomass sloughing. This equation is as follows:

$$C_n - 1 = \frac{\sqrt{1 + 4kt(C_n - 1)}}{2kt} \qquad (4.23)$$

where:
- C_n = is the concentration of soluble organics in the nth stage (mg/L)
- k = is the second-order reaction constant of 0.083 (L/mg/hr)
- t = is the average hydraulic residence time in the nth stage (hour)
- C_n-1 = is the concentration of soluble organic matters entering the ninth stage (mg/L).

The design engineer needs to be aware that this equation may be used only where appropriate, and that there may be a number of applicable equations in the available RBC literature.

Design safety factor

Effluent concentrations of ammonia nitrogen from the RBC process designed for nitrification are affected by diurnal load variations. An evaluation of equalization vs. additional RBC media surface area is required when consistently low ammonia nitrogen levels are necessary to meet effluent limitations. If flow equalization is not provided, then it may be necessary to increase the design surface area proportional to the ammonia nitrogen diurnal peaking rates.

Secondary clarification (SC)

The concentration of suspended solids leaving the last stage of an RBC system treating municipal wastewater is generally less than 200 mg/L when preceded with primary clarification. To attain secondary effluent quality standards, secondary clarifiers must be used in conjunction with RBCs. The surface overflow rate, generally, should not exceed 32.6 m^3/m^2·day (800 gallons per day per square foot) for secondary clarifiers. Consideration may be given to covering the clarifiers to improve efficiency.

4.7 Combined aerobic processes

The combined aerobic processes are designed to sustain shock loads in activated sludge, aerobic contactor, and trickling filter (biofilter) processes. There are several possible combinations of these aerobic processes; activated biofilter process; trickling filter solids-contact process; biofilter activated-sludge process; and trickling filter-activated sludge process.

The activated biofilter process is a trickling filter with recycling of secondary sludge back to the trickling filter, to create a higher level of BOD removal through a combined attached and suspended microbial growth. High BOD loading in wastewater streams such as those found in food processing operations can be amply handled. Design loading of BOD normally ranges from 3.21 to 4.0 kg/m^3·day for 60–65 % BOD removal in the filter (Arora & Umphres, 1987).

The basic trickling-filter solids-contact process consists of a trickling filter, an aerobic contact tank, and a final clarifier. The trickling filter does the most BOD removal (about 65–85%), according to Parker *et al*. (1993). The biosolids formed on the filter are sloughed off and concentrated through sludge recirculation in the contact tank. In the contact tank, the suspended solids are aerated for less than one hour, causing flocculation of the solids and thus further removing BOD. The overall performance of a TF/SC process is determined by calculating the BOD removal from the trickling filter and the aerobic contact tank (Parker & Bratby, 2001).

4.8 Contact anaerobic systems

Anaerobic treatment is used in both biological wastewater treatment and sludge digestion. Anaerobic treatment using contact anaerobic systems is an effective treatment method for treating high-strength wastewater containing large amounts of organic materials (high BOD). Food and agricultural wastewater sometimes falls into this category of wastewater streams (such as blood water or stick-water). For example, meat-packing and fish-processing wastewater are treated successfully with anaerobic treatment processes. In the pilot scale studies of anaerobic treatment of fish processing wastewaters, the removal rates of solids were 75–80%, with loads ranging from 3–4 kg COD/m^3·day of digester (Balslev-Olsen *et al*., 1990; Mendez *et al*, 1990).

The microbiology of the anaerobic treatment involves facultative and anaerobic microorganisms which, in the absence of oxygen, convert organic materials into gaseous methane and carbon dioxide. The anaerobic process consists of two distinct stages: acid fermentation and methane fermentation. The process starts with degrading insoluble complex organic materials such as proteins into soluble organic materials which, in turn, are consumed by acid, producing bacteria to yield volatile fatty acids, along with carbon dioxide and hydrogen. The methane-producing bacteria devour precedent biochemical products to produce methane and carbon dioxide. Figure 4.7 summarizes the reactions and intermediates involved in an anaerobic treatment.

4.8.1 Advantages and disadvantages of anaerobic processes

Compared with aerobic wastewater treatment processes, anaerobic processes have certain advantages:

Figure 3.3 Photograph of an inclined screen

Figure 3.5 Schematic diagram of a dissolved air flotation system

Food and Agricultural Wastewater Utilization and Treatment, Second Edition. Sean X. Liu.
© 2014 John Wiley & Sons, Ltd. Published 2014 by John Wiley & Sons, Ltd.

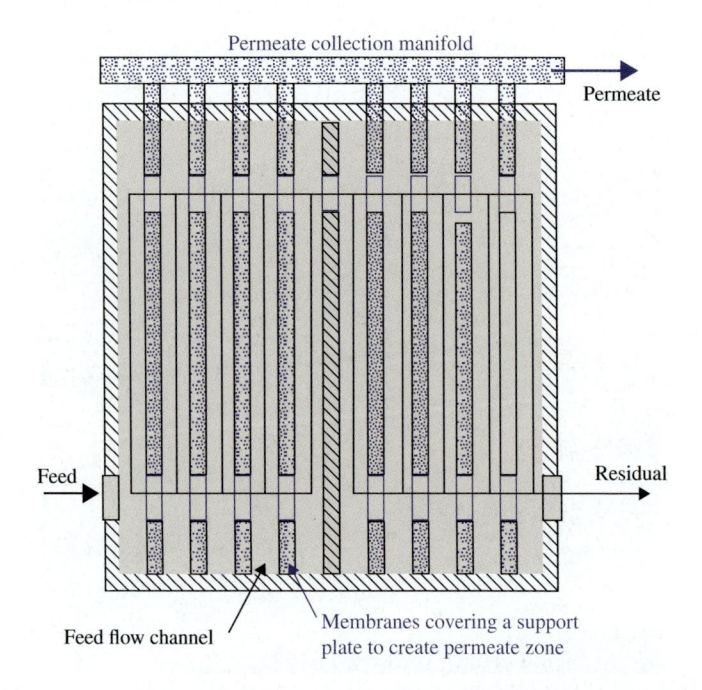

Figure 3.16 A schematic illustration of a plate and frame module (Courtesy of Dr. Leland Vane of US EPA)

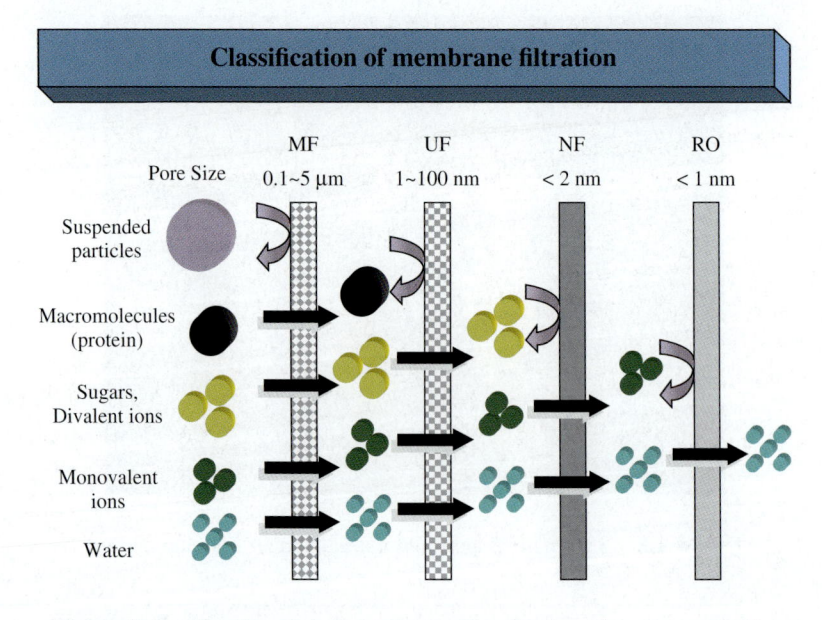

Figure 5.4 Classification of membrane filtration based on size exclusion

Figure 6.1 A waste stabilization pond system in Mèza, France. It contains two small anaerobic ponds, four experimental facultative ponds, and one series pond system composed of a facultative pond and four maturation ponds

Figure 6.4 Photograph of an anaerobic pond for wastewater treatment in the field

Figure 6.10 A surface flow wetland system in the USA. *Source*: USEPA

Figure 7.8 A freezing-thawing sludge bed

Figure 7.9 A sample of freezing-and-thawing treated sludge from anaerobic digester

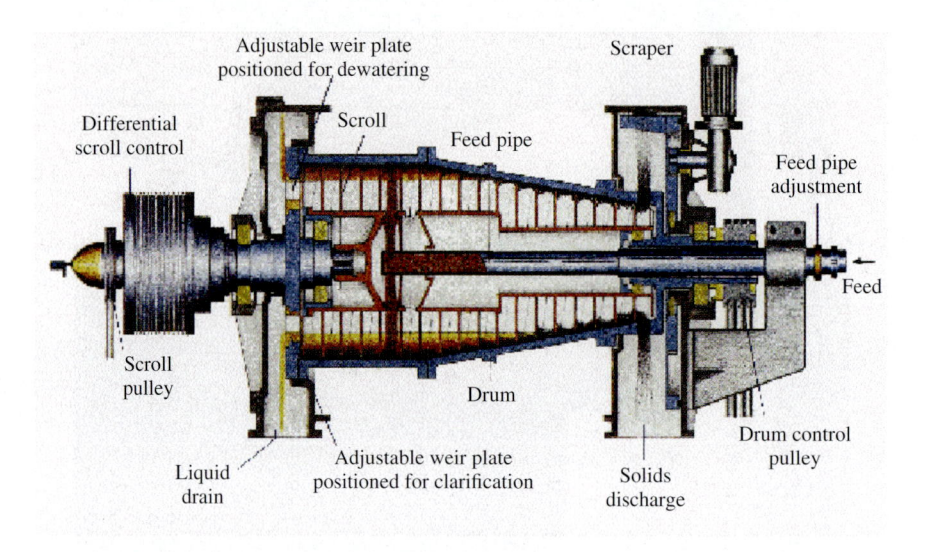

Figure 7.6 A schematic illustration of centrifuge for dewatering of sludge

Figure 7.7 A filter press for sludge dewatering

Figure 7.3 Photo of a reed bed sludge treatment before planting vegetation

Figure 7.4 Photo of a reed bed sludge treatment after planting vegetation

Figure 6.14 Three species of native duckweeds in Florida waterway. *Source*: Willey Durden, USDA ARS (retired)

Figure 7.2 Aerial photo of Gild Bar Wastewater Treatment Plant in Edmonton, Alberta, Canada

Figure 6.13 Water hyacinth in Florida waterway. *Source*: Willey Durden, USDA ARS (retired)

Figure 6.11 A subsurface flow wetland system in the US. *Source*: USEPA

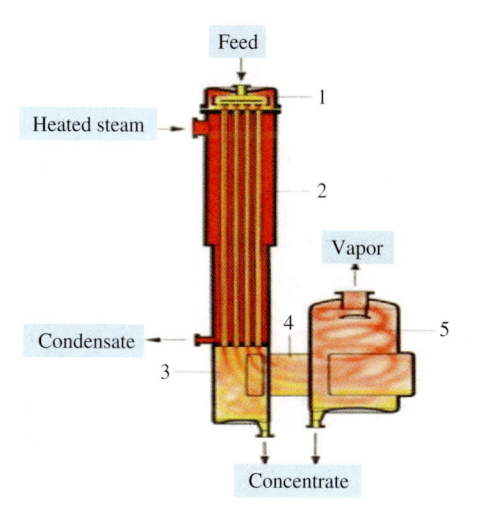

Figure 7.13 Schematic diagram of multiple-effect evaporator for sludge dewatering

Figure 8.1 Sludge from anaerobic digestion of potato processing wastewater. *Source*: Steven Vaughn

Figure 8.2 Cavendish biogas facility. *Source*: Steven Vaughn

Figure 8.3 Red Robin tomato plants grew in different substrates after four weeks. *Source*: Steven Vaughn

4.8 CONTACT ANAEROBIC SYSTEMS

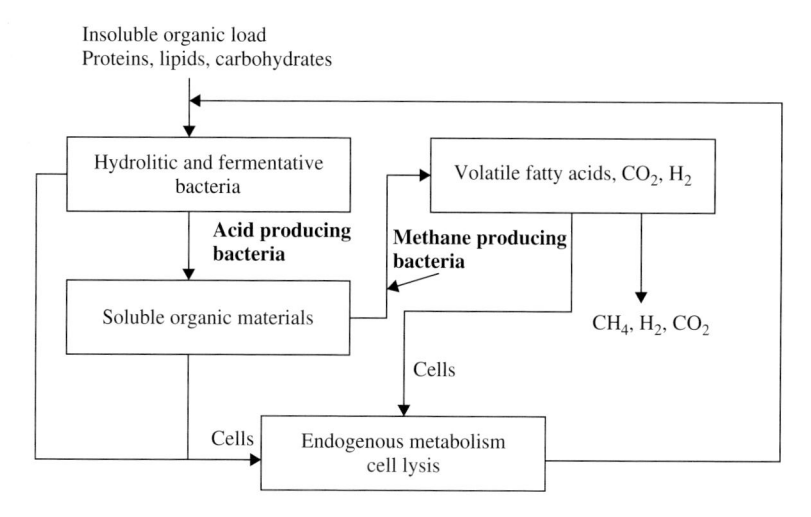

Figure 4.7 Diagram of the reactions and intermediates involved in an anaerobic treatment

- Biomass produced with anaerobic processes is much lower, thus reducing costs associated with sludge treatment and management.
- Treatment of high-strength, organic-rich wastewater is better with anaerobic processes, because anaerobic processes are not limited by oxygen transfer rate (usually the bottleneck of aerobic processes).

However, anaerobic processes of wastewater treatment also suffer from the following shortcomings:

- Higher thermal energy is required to maintain temperature needed for anaerobic processes; however, this problem can be overcome with the possible utilization of methane from the processes.
- Higher holding time (or detention time) for completing the processes.
- Undesirable odors are commonly associated with anaerobic processes due to the formation of hydrogen sulfide and mercaptans. Certain food wastewater sources are rich in sulfur and nitrogen compounds, which could aggravate the odor problem.
- Sludge from anaerobic processes is harder to treat, thus requiring extra money/time/equipment to handle it.
- Anaerobic systems are difficult to operate and subject to negative effect of "shock loading".

Table 4.2 Common process and performance data for anaerobic processes used for the treatment of industrial wastewaters

Process	Input COD, mg/L	Hydraulic detention time, h	Organic loading kg COD/m³·d	COD removal %
Anaerobic contact process	1,500–5,000	2–10	0.001873–0.009364	75–90
Upflow anaerobic Sludge–blanket	5,000–15,000	4–12	0.015607–0.078035	75–85
Fixed–bed	10,000–20,000	24–48	0.003746–0.01873	75–85
Expanded–bed	5,000–10,000	5–10	0.01873–0.03746	80–85

Source: Metcalf & Eddy (1991).

4.8.2 Anaerobic contact processes

The anaerobic wastewater treatment processes in use include the anaerobic contact process, the upflow sludge-blanket reactor, the anaerobic filter (fixed bed) reactor, and the expanded-bed process. The loading and performance data for anaerobic contact processes are summarized in Table 4.2.

Upflow anaerobic sludge-blanket process (UASB)

The upflow anaerobic sludge-blanket process is a suspended-growth bioreactor and its schematic diagram is shown in Figure 4.8. As its name indicates, the wastewater flows in from the bottom of the tank upwards, and active anaerobic bacteria convert the waste into methane and carbon dioxide. A sludge blanket is developed in the lower portion of the tank, and the component particles in the sludge blanket are aggregated to resist the hydraulic shear of the upwardly flowing wastewater and prevent the blanket from being carried over and out of the tank.

In general, the UASB process is capable of achieving high removing efficiency at high COD loadings. Lettinga *et al.*, (1980) conducted anaerobic treatment of sugar beet wastewater using UASB systems at different scales. The experiments, carried out in a 6 m³ pilot plant, had been shown to be capable of handling organic space loads of 15–40 kg COD/m³ ·day at 3–8 hour liquid detention times. In the first 200 m³ full-scale plant of the UASB concept, organic loadings of up to 16 kg COD·m⁻³/day could be treated satisfactorily at a detention time of four hours. The performance of UASB reactors was limited by the capability of the gas–liquid separator to withhold the sludge in the reactors.

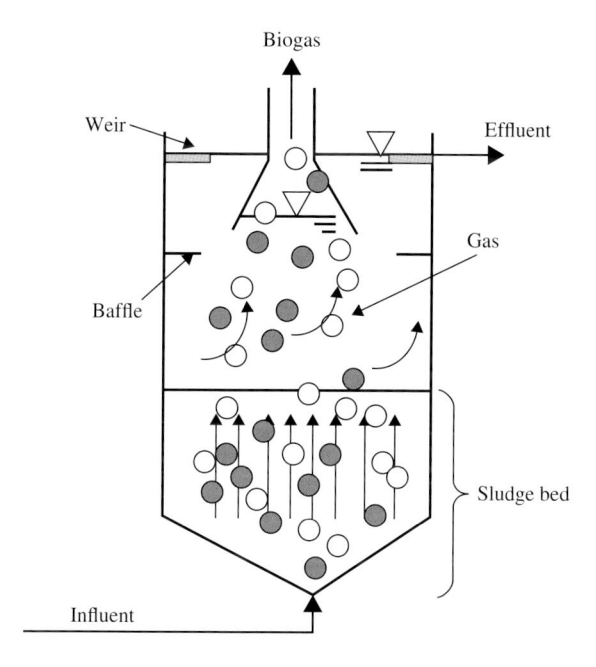

Figure 4.8 Schematic diagram of upflow anaerobic sludge-blanket process

The design of an UASB reactor needs to ensure an adequate sludge zone, as most sludge resides there. Sometimes baffles are added to the locations above the sludge blanket to assist in separating biogas, sludge, and liquid.

Anaerobic filter (AF)

In a similar configuration to that used with a trickling filter, an anaerobic filter reactor uses filter media to support anaerobic microorganisms to degrade carbonaceous organic matter in the wastewater feed from the bottom of the reactor (upflow mode) or from the top of the reactor (downflow mode), as shown in Figure 4.9. Since the bacteria are retained on the media in the column, the residence time of these cells are quite high, which allows bacteria to have sufficient time to remove the organic matter. This long residence time also enables the anaerobic filter to adapt to varied operating conditions. The fluid dynamics of AF reactors is either completely mixed or partially mixed, depending on the rate of recirculation. The filter medium bed is subject to bed clogging, due to accumulation of biological and inorganic solids; however, periodical backwashing will alleviate the clogging problem and associated head loss.

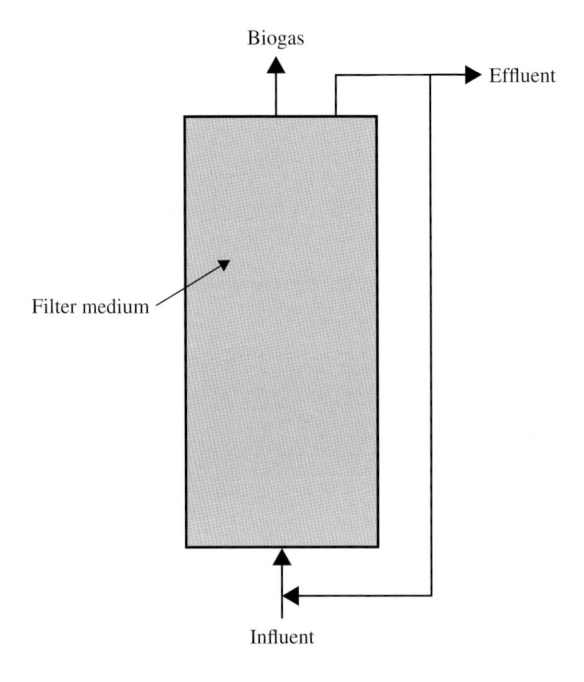

Figure 4.9 Schematic diagram of upflow mode anaerobic filter (AF)

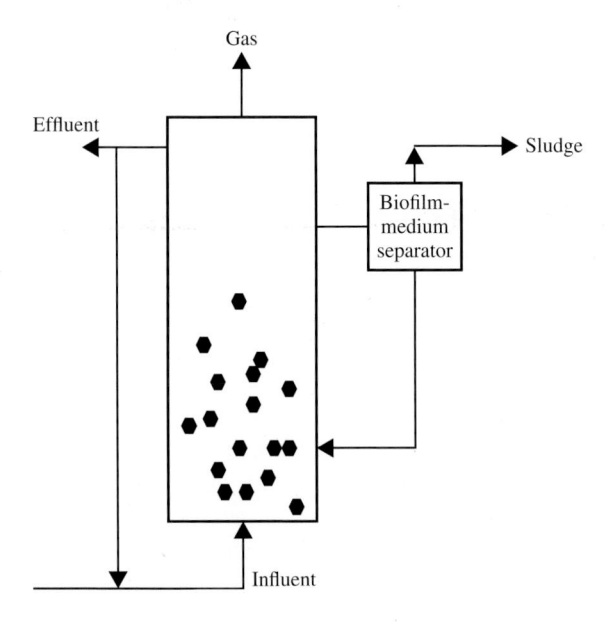

Figure 4.10 Schematic diagram of anaerobic fluidized bed reactor (AFBR)

Anaerobic fluidized bed reactor (AFBR)

The anaerobic fluidized bed reactor is an expanded-bed reactor (Figure 4.10) that is filled with a solid medium used for hosting anaerobic bacteria. The wastewater is fed from the bottom of the reactor and flows upward through the medium-containing column of the reactor, which is able to retain the media in the suspension from drag forces exerted by the upflowing wastewater stream.

The effluent from the reactor is recycled to dilute the influent and to maintain an adequate flow rate, so that the bed remains expanded (fluidized). Due to the expansion of the medium bed, substantially larger amount of biomass can be maintained without incurring the bed clogging and subsequent head loss incurred with anaerobic filter reactors. Reports of biomass concentration of 15,000 to 40,000 mg/L in AFBR reactors are common in the literature (Metcalf & Eddy, 1991). Because of high flow rate maintained in the AFBR reactors, especially for treating high-strength food wastewater, AFBR reactors can be designed to behave like an ideal completely mixed reactor, a CSTR.

The choice of medium in AFBR needs to be further explained; the ideal medium must be light (easily fluidized), small (easily fluidized and high surface-to-volume ratio), porous with large voids (more space for biological and inert solids), inert (to chemical and biological reactions), and resistant (to abrasion and erosion). Silica sand, anthracite coal, reticulated polyester foam, and activated carbon are common media types used in AFBR reactors.

4.9 Further reading

Grady, C.P.L. & Lim, H.C. (1980). *Biological Wastewater Treatment: Theory and Applications*. Marcel Dekker, New York, NY.

Metcalf & Eddy, Inc. (1991a). *Wastewater Engineering, Treatment, Disposal, and Reuse. Third edition*. McGraw-Hill, New York, NY.

4.10 References

Arora, M.L. & Umphres, M.B. (1987). Evaluation of activated biofiltration and activated biofiltration/activated sludge technologies. *Journal – Water Pollution Control Federation* **59**, 4.

Balslev-Olsen, P., Lynggaard, A. & Nickelsen, C. (1990). Pilot-plant experiments on anaerobic treatment of wastewater from a fish processing plant. *Water Science and Technology* **22**, 463–474.

Lettinga, G., van Velsen, A.F.M., Hobma, S.W., de Zeeuw, W. & Klapwijk, A. (1980). Use of the upflow sludge blanket (USB) reactor concept for biological wastewater treatment, especially for anaerobic treatment. *Biotechnology and Bioengineering* **22**, 699–734.

Mendez, R., Omil, F., Soto, M. & Lema, J.M. (1990). Pilot plant studies on the anaerobic treatment of different wastewaters from a fishing canning factory. *Water Science and Technology* **25**, 37–44.

Metcalf & Eddy (1991b). *Wastewater Engineering, Treatment, Disposal, and Reuse. Third edition*. McGraw-Hill, New York, NY.

Nguyen, V.T. & Shieh, W.K. (2000). Secondary treatment. In: Liu, D.H.F. & Liptak, B.G. (eds.) *Wastewater Treatment*. Lewis Publishers, New York, NY.

Opatken E.J. (1986). An alternative RBC design-second order kinetics. *Environmental Progress* **5**, 51–56.

Parker, D.S. & Bratby, J.R. (2001). Review of two decades of experience with TF/SC process. *Journal of Environmental Engineering, ASCE* **127**, 380–387.

Parker, D.S., Brischike, K.V. & Matasci, R.N. (1993). Upgrading biological filter effluents using the TF/SC process. *Journal of the Institution of Water & Environmental Management* **7**, 90–100.

5
Advanced wastewater treatment processes

5.1 Introduction

Biological treatment processes, in combination with primary treatment (sedimentation), typically remove 85% of the BOD_5 and soluble solids originally present in the raw wastewater and some of inorganic materials. As a treatment process, activated sludge systems generally produce an effluent of slightly higher quality, in terms of its constituents, than trickling filters or RBCs. When coupled with a disinfection step, all these processes can provide substantial (although not complete) removal of bacteria and viruses. However, they remove very little phosphorus, nitrogen, non-biodegradable organics, or dissolved minerals and, in an increasing number of cases, this level of treatments has proved to be insufficient to protect the receiving waters from contamination or to provide reusable water for industrial recycle.

As a consequence, additional treatment steps have been added to wastewater treatment plants to provide for further organic and solids removal, or to provide for the removal of nutrients and/or toxic materials. These post-primary and/or post-secondary wastewater treatment processes are grouped into a category of treatment schemes called advanced wastewater treatment. In a way, advanced wastewater treatment can be defined broadly as any process designed to produce an effluent of higher quality than that normally achieved by secondary treatment processes, or containing unit operations not normally found in secondary wastewater treatment facilities.

The treatment processes in advanced wastewater treatment can be simply a number of unit operations added to the existing primary and/or secondary treatment processes (sometimes, this arrangement is called tertiary treatment), or may be entirely stand-alone units of physicochemical processes, biological processes, or a combination of the above. The specific processes employed in advanced wastewater treatment may be similar to those used in primary or/and

Food and Agricultural Wastewater Utilization and Treatment, Second Edition. Sean X. Liu.
© 2014 John Wiley & Sons, Ltd. Published 2014 by John Wiley & Sons, Ltd.

secondary wastewater treatment, or they may be totally different from those in conventional wastewater treatment processes, in order to remove the pollutants that cannot either be removed at all or be reduced to the quantity that is at the safe level. The applications of advanced wastewater treatment depend on the treatment goal of the effluents. Whether it is to remove excessive amount of BOD/solids, nutrients or heavy metals/toxic materials, advanced wastewater treatment has an important role in ensuring the quality of treated wastewater.

The pollution problems arising from excessive amount of certain nutrients in wastewater are the most common reason for advanced wastewater treatment, and they are mainly caused by nitrogen-rich and phosphorous-rich compounds. Nitrogen-rich substances, such as proteins, are biologically converted into ammonia through a process called ammonification.

An excessive amount of nutrients in discharged treated wastewater will lead to a problem called eutrophication. The eutrophication problem associated with algal blooms and de-aeration of the receiving water body results from the oxidation of ammonia to nitrate by the nitrifying bacteria (a process called nitrification), and this can suffocate fishes and other animals living in the polluted water. High concentrations of nitrate in water are toxic both to humans and to animals. A schematic diagram in Figure 5.1 shows the simplified nitrogen cycle in the environment.

Another area of pollution problem associated with wastewaters that is not addressed adequately by the conventional wastewater treatment is a category of priority pollutants and volatile organic compounds (VOCs), which has been identified by the regulatory agencies since the early 1980s. These pollutants are toxic to humans and the aquatic ecosystem; if left untreated, they will enter the domestic water supply when insufficiently treated wastewater is allowed to discharge to surface water or groundwater.

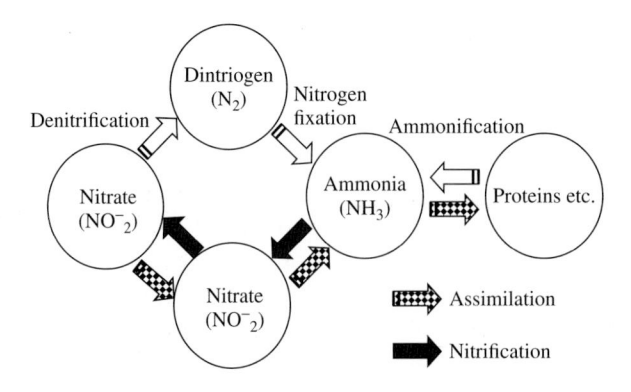

Figure 5.1 Simplified diagram of nitrogen cycle in nature

There are a number of physicochemical methods useful for achieving nitrogen removal from wastewater effluents, such as ammonia air stripping and ion exchange. However, the high cost and low reliability of these methods have hindered their popularity. By far the most widely used methods of nitrogen removal in practice have been biological processes. Phosphorus can be removed by both biological and physiochemical processes.

Recent interest in combining removal of nitrogen and phosphorus in agricultural and food wastewaters with bioenergy production has started transforming the landscape of advanced treatment of agricultural and food wastewater (see Chapter 8 for more on this new, exciting development). VOCs are found to be quite resistant to biological treatment, while physiochemical processes such as air stripping have their own limitations in terms of VOC removal from wastewater and may turn a water pollution problem into an air pollution quandary. Membrane technology, particularly pervaporation, has now grown from being perceived as an academic laboratory curiosity, as it once was, and has found practical uses in various wastewater treatment projects (Peng et al., 2003).

5.2 Biological removal of nitrogen: nitrification and denitrification

The goal of nitrogen removal, regardless of exactly which nitrogen compounds are actually present in wastewater streams, has been the production of nitrogen gas – a relatively inert, water-insoluble gas that is easily separated from liquid media. The necessity of producing nitrogen gas in the treatment processes of nitrogen removal is mainly due to the high solubility of nitrogen compounds such as NO_3^-, NH_4^+, and NO_2^- that may be present. However, there is some indication that this old paradigm is being challenged. Since nitrogen gas produced in wastewater treatment operations does not have economical value, some researchers are seeking to remove nitrogen compounds in dissolved forms (Aiyuk et al., 2004). The most promising method of removing dissolved nitrogen compounds is the application of adsorption, employing zeolite columns in an integrated wastewater treatment process. The recovered nitrogen compounds can be used as fertilizers. However, due to the high cost of zeolite columns, the most economical way of removing nitrogen compounds from wastewater streams at present is to employ conventional biological processes, consisting of nitrification and denitrification.

The biology of nitrification-denitrification has been briefly discussed in Chapter 2; in a nutshell, biological nitrogen removal from wastewater involves converting organic nitrogenous compounds to ammonia, then nitrate (nitrite), and finally to gaseous nitrogen, as illustrated in the nitrogen cycle diagram in Figure 5.1. Organic nitrogen materials in food and agricultural wastewater steams are either in the forms of proteins, nucleic acids and urea, or as ammonium ions (NH_4^+). Normally, domestic and most industrial wastewater streams rarely contain nitrate or nitrogen. Nitrate in some agricultural wastewater streams may come from fields in which excessive amounts of nitrogen-rich man-made fertilizers have been applied.

Nitrification is a microbial process by which reduced nitrogen compounds (primarily ammonia) are sequentially oxidized to nitrite, and nitrate and is the first step in the removal of nitrogen from wastewater streams by the nitrification-denitrification process. It is primarily accomplished by two groups of autotrophic nitrifying bacteria, *Nitrosomonas* and *Nitrobacter*, which can build organic molecules using energy obtained from inorganic sources – in this case ammonia or nitrite. In the first step of nitrification, ammonia-oxidizing bacteria oxidize ammonia to nitrite. *Nitrosomonas* oxides ammonia to the intermediate product, nitrite and nitrite is further converted into nitrated by *Nitrobacter*. Overall equations for nitrite production and nitrate formation by these two categories of nitrifying bacteria are represented in Equations (2.9) and (2.10) (see Chapter 2).

A number of environmental factors can influence the nitrification process, including substrate concentration, temperature, oxygen, pH, and toxic or inhibiting substances. Nitrifying bacteria are susceptible to a number of inhibitors, both organic and inorganic agents. They are also sensitive to pH value, and a range of 7.5 to 8.6 is found to be optimal for the growth of nitrifying bacteria. There is also a dissolved oxygen level that could limit the nitrification process; a concentration of above 1 mg/l is essential for nitrification to occur.

Temperature has a strong effect on nitrifying bacteria, just as in the case of heterotrophic aerobic bacteria. The temperature dependence for the nitrification process fits an Arrhenius type of equations, at least lower than 30°C. At higher temperatures (30–35°C), the growth rate of nitrifying bacteria is constant, and it starts declining between 35°C and 40°C (Henze *et al.*, 2001).

5.3 Physicochemical removal of nitrogen

As described previously, biological nitrogen removal is not the only technology available for removal of nitrogen compounds from wastewaters. Some non-biological processes are able to recover nitrogen compounds in their dissolved forms for potential uses as fertilizers, and these can be viable alternatives under some circumstances. On the whole, however, physicochemical processes for removing nitrogen from wastewater are not practically popular. The reasons for their unpopularity are often cited as cost, inconsistent performance, and operating and maintenance problems. The principal processes employed for nitrogen removal from wastewaters are air stripping, breakpoint chlorination, and selective ion exchange.

Air stripping is used to remove ammonia from wastewater, because ammonia in water can be easily volatilized and carried away by flowing air stream, particularly in high pH values. The gaseous ammonia in air can later be captured. The process is temperature sensitive, and fogging and icing occur in cold temperatures. This is an expensive operation that requires lime for pH control, and that can only be justified in some special cases, such as the need for a high pH for other reasons at the time.

Breakpoint chlorination is a process that involves the addition of chlorine to wastewater to oxidize the ammonia in the wastewater to nitrogen gas and other possible stable compounds., Theoretically, with proper control, it can remove all ammonia in wastewater, and it is therefore used to "polish" the effluents from other nitrogen removal processes. The downside of the breakpoint chlorination process is that it is sensitive to pH, thus requiring proper control of pH with skillful operators during the application of chlorine. The process is expensive because of the use of chlorine and skilled labor requirements, and its residue may be toxic to the aquatic life.

Ion exchange is a separation technology (see Chapter 3 for the description of ion exchange systems) that utilizes ion-selective resins (mostly synthetic, but there are also natural alternatives available, such as zeolites) to remove certain ions from wastewater. In ion exchange operations, the ions in wastewater displace the ions on the resins, thus separating from the rest of the components in the wastewater streams. The removed ions in the resins can be washed out with appropriate regenerants. For nitrogen removal with a zeolite ion exchange system, lime can be used to regenerate the zeolites. Because of the presence of organic materials in wastewater, restorants such as sodium hydroxide, hydrochloric acid, methanol and bentonite are employed to remove organic materials from resins.

5.4 Biological removal of phosphorus

Phosphorus is a constituent of wastewater, averaging around 10 mg/L in most cases, and the principal form in food and agricultural wastewater is organically bound phosphorus. Organically bound phosphorus originates from body and food waste which, upon biological decomposition of these solids, is converted to orthophosphates.

Biological phosphate removal is a relative new technology that dates back to the late 1950s, but it was not until the 1970s that there were full-scale processes developed for practical use in advanced wastewater treatment. Based on a series of tests and experiments on biological phosphorus removal, Fuhs & Chen (1975) had determined that a genus called *Acinetobacter* was responsible for biological phosphorus removal, and they postulated that these bacteria utilized substrates, a type of volatile fatty acids produced from anaerobic phase, for growth and excessive phosphorus uptake under aerobic conditions. However, this explanation of biological phosphorus removal was challenged by a number of researchers using molecular ecology techniques (e.g., Bond *et al.*, 1995; Mino *et al.*, 1998). Nevertheless, it was recognized from the process engineering point of view that the necessary condition for biological phosphorus removal is the existence of a true anaerobic phase in the process. This insight has helped the development of several process configurations of biological phosphorus removal throughout the world.

It was later discovered that the main function of the anaerobic phase was not only to provide polyphosphate-accumulating bacteria with volatile fatty acids (VFAs), but also to enable this type of bacteria to use phosphate as an energy reserve to pick up substrates (Wentzel *et al.*, 1986; Arun *et al.*, 1987; Smolders *et al.*, 1994; Maurer *et al.*, 1997; Mino *et al.*, 1998). The available VFAs enable bacteria to utilize the VFAs as a carbon source under anaerobic conditions and to release phosphate into the solution. A subsequent aerobic phase, as well as anoxic conditions, take up the phosphate in the water and, as a result, a greater amount of phosphorus is removed from wastewater as sludge. Glycogen is also utilized under anaerobic conditions and is replenished during the aerobic phase of the cycle.

It is clear today that *Acinetobacter* is responsible only for some portions of biological phosphorus removal, and that phosphorus-accumulating organisms (PAOs) include many heterotrophic microorganisms. However, not all heterotrophic bacteria are PAOs and, in a wastewater treatment plant with biological phosphorus removal processes, these non-PAOs may compete with the heterotrophic PAOs for the substrate, particularly those low molecular fatty acids that are needed for the phosphorus storage mechanism. The result of this competition determines how successful is the biological phosphorus removal process. As mentioned previously, the anaerobic phase is of great importance in steering the substrate utilization towards the direction of the heterotrophic PAOs.

The biological treatment or removal of phosphorus from wastewater depends on the accumulation of a large amount of bacteria that are capable of storing phosphorus in the form of polyphosphate inside the bacterial cells. Polyphosphate, as stored energy for bacteria, is produced as a result

5.4 BIOLOGICAL REMOVAL OF PHOSPHORUS

of sequestering of volatile fatty acids by aerobic bacteria under anaerobic conditions, resulting in poly(hydroxyalkanoates)(PHAs) under simultaneous use of glycogen. This requires that the influent of wastewater for biological phosphorus removal must first mix with sludge in order to create a true anaerobic environment, free from electron acceptors such as oxygen and nitrate. In the anaerobic environment or zone, volatile fatty acids (which may be formed by fermentation) in the incoming wastewater stream can be accumulated by polyphosphate-accumulating bacteria. Thus, the successful design of a biological phosphorus removal process relies on the creation of such a true anaerobic zone, and will also be influenced by the characteristics of incoming wastewater streams.

Depending on whether there is a presence of volatile fatty acids produced by fermentation, the size of the anaerobic zone or reactor varies with the predominant anaerobic process. The volatile fatty acids-containing wastewater streams require small reactors, while incoming wastewaters from an aerobic process without volatile fatty acids will need larger reactors because the anaerobic phase has to be based on slower fermentation process. Figure 5.2 shows a basic schematic diagram of a biological phosphorus removal process. In this diagram, substrate is taken up by polyphosphate-accumulating bacteria and phosphate is released into the liquid phase in the anaerobic phase; in the aerobic phase, the bacteria grow and accumulate phosphate in the cells, resulting in removal of phosphate from the wastewater.

It is generally observed that, in order to operate a successful biological phosphorus removal process, it is crucial that the incoming wastewater stream should contain the correct balance of nutrients, carbon sources, and pH. Careful considerations must be given to the food/microorganism ratio, hydraulic retention time, solid retention time, temperature, and DO concentration (Mulkerrins *et al.*, 2004). In addition to this basic biological phosphorus removal process,

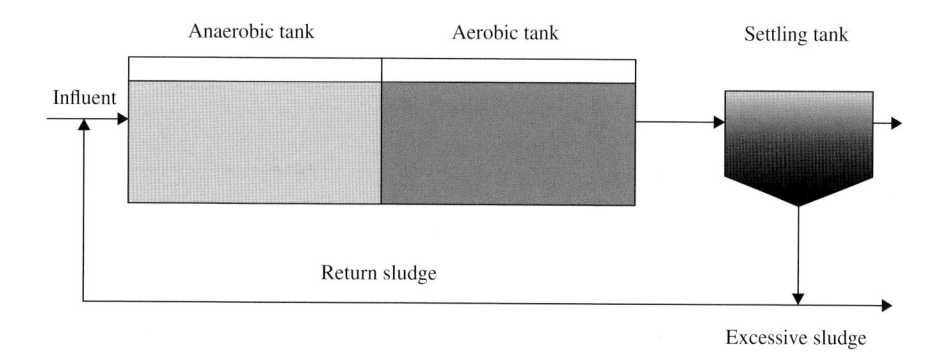

Figure 5.2 Schematic diagram of a typical biological phosphorus removal process

there are several combined processes of chemical and biological phosphorus removal. Figure 5.3 shows a schematic diagram of metabolisms of PAOs under aerobic and anaerobic conditions.

It has been shown that some PAOs, called denitrifying PAOs, can also accomplish denitrification while accumulating phosphorus. The combined processes of biological phosphorus and nitrogen removal by denitrifying phosphate-accumulating bacteria such as UCT (University of Cape Town)-type processes have been demonstrated (Kuba *et al.*, 1993, 1994). In these combined processes, PAOs used nitrate or nitrite as the electron acceptor instead of oxygen. Contrary to earlier view of different PAOs involved in denitrification and biological phosphorus removal in the combined processes, it now appears that *Accumulibacter* was the denitrifying PAOs in both anaerobic and anoxic conditions in the combined processes (Ahn *et al.*, 2002; Zeng *et al.*, 2003). Further microbial analysis in the study of Zeng *et al.* (2003) revealed that *Accumulibacter* was the dominant species in both PAO and Denitrifying PAO sludge. The current trend in biological phosphorus removal is development of a simultaneous nitrification, denitrification, and phosphorus removal process that can save capital and operational costs, as well as having a smaller environmental footprint.

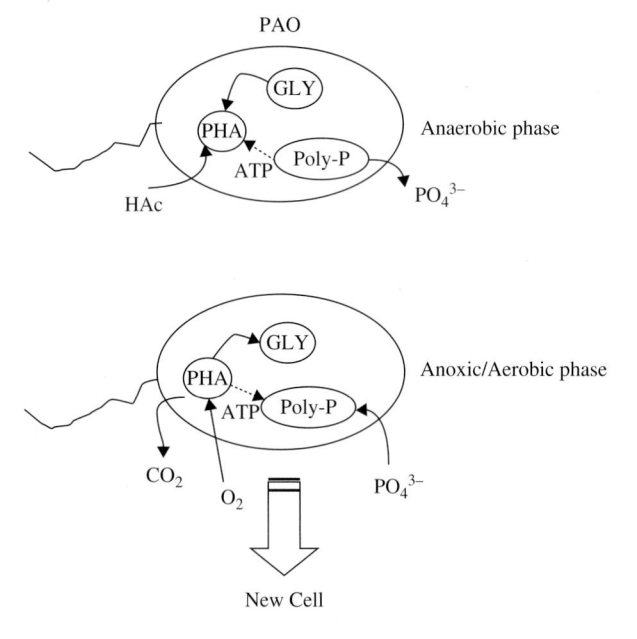

Figure 5.3 Schematic diagram of metabolisms of PAOs under anaerobic and aerobic conditions

5.5 Physicochemical removal of phosphate

There are several physicochemical phosphate removal processes that can be used with conventional secondary wastewater treatment. These processes generally involve using chemicals to facilitate precipitation of phosphate, and a primary clarifier (sedimentation tank or basin) to separate phosphate-containing sludge from the treated wastewater. Alternatively, in some cases, dissolved air flotation is used to remove the phosphorus-containing complexes. The principal chemicals employed in this type of removal are alum, sodium alimunate, ferrous chloride or sulfate, ferric chloride or sulfate, and lime. Ferrous sulfate and ferrous chloride are available as by-products of steel-making operations (called pickle liquor) and, thus, have cost advantages in use. Polyelectrolytes have been used effectively along with alum or lime as flocculant aids. The assortment of chemicals used in the precipitation of phosphate has a striking similarity of those flocculants employed in the formation of flocs from colloidal dispersion of particulates.

The chemistry of chemical precipitation reactions involved may be reviewed from the materials described in Chapter 3. The choice of using any of the chemicals mentioned above for phosphorus removal is determined by the following factors:

- incoming flow phosphorus level;
- suspended solids and colloids;
- alkalinity;
- chemical cost including transportation;
- reliability of chemical availability;
- sludge handling facilities and method as well as cost including disposal;
- compatibilities with other treatment processes (primary and secondary treatment processes).

Iron and alum salts can be added at various points in primary and secondary treatment processes. However, in order to achieve maximum removal of phosphorus, alum and iron salts are best added after the secondary treatment in organic phosphorus-containing wastewater streams (where organic phosphorus is transformed as orthophosphorus). Additional nitrogen removal might occur as a result of this sequence of adding alum or iron salts. The various chemical addition points and sequences are illustrated in Figure 3.1 (Chapter 3).

The obvious drawback of physicochemical processes of phosphorus removal is the cost associated with the chemicals or coagulants. Additionally, the increase in sludge volume due to the addition of chemicals and inability of the physicochemical processes to remove nitrogen compounds is another known disadvantage.

5.5.1 Lime precipitation of phosphate

A lime phosphate precipitation system is similar to a primary settling tank or basin. Wastewater inflow is mixed with lime and flocculated to cause the precipitation of phosphate in the wastewater in the form of insoluble calcium salt. The main reaction required to determine quantity of sludge produced during the precipitation of phosphorus with lime is expressed as:

$$5Ca^{2+} + 3PO_4^{3-} + OH^- \leftrightarrow Ca_5(PO_4)_3(OH)\downarrow \tag{5.1}$$

However, the same lime may also cause precipitations of $Mg(OH)_2$ and $CaCO_3$ in addition to orthophosphorus.

The lime sludge from the clarifier is often recycled in order to reduce the cost of lime in the treatment. The process generally achieves 80–95% removal rate of phosphate in wastewater. Lime can also soften the water, but using large amounts of lime presents a new challenge – dealing with the increased amount of sludge in the treatment.

5.5.2 Ferrous precipitation of phosphate

This process uses ferrous chloride to convert soluble phosphate in wastewater streams into an insoluble form. The relevant reaction of phosphorus precipitation in presence of ferrous salts can be written as:

$$3Fe^{2+} + 2PO_4^{3+} \leftrightarrow Fe_3(PO_4)_2\downarrow \tag{5.2}$$

It is possible that $Fe(OH)_2$ may also be generated and precipitate out when ferrous salts are added.

Polyelectrolyte is also used to aid flocculation of phosphate complex with ferrous chloride in the clarifier. Other salts can also be used in lieu of ferrous chloride, such as ferric chloride and sodium aluminate. Recall that in Chapter 3 we discussed flocculation using iron salts and polyelectrolytes to reduce colloidal particulates. This process can be viewed as an extension to the conventional primary wastewater treatment processes.

Like all processes involving chemical precipitation reaction, the cost of ferrous precipitation of phosphate depends on the cost of metal salts. If the metal salts can be obtained economically, the cost of the phosphate removal operation will be reasonable. Ferrous chloride used in wastewater treatment can be obtained from steel plants, in which ferrous chloride is a waste product of steel production.

Application of ferrous chloride to the primary sedimentation facilities for phosphate removal also has unintended benefit; it reduces the BOD load on secondary treatment facilities, because it causes flocculation of, and subsequently settling of, organic colloids from wastewater. The downside of the use of ferrous chloride as flocculants is that it requires the use of corrosion-resistant equipment upon which the ferrous or ferric sludge cannot inflict damage. The sludge may also not be compatible with some of the dewatering techniques described in Chapter 6.

5.5.3 Alum and ferric precipitation of phosphate

A representative reaction expression for phosphorus removal with addition of alum is as follows:

$$Al^{3+} + PO_4^{3-} \leftrightarrow AlPO_4 \downarrow \quad (5.3)$$

It is likely that $Al(OH)_3$ may also precipitate from the wastewater.

A pertinent reaction for phosphorus removal with addition of ferric salts is expressed as:

$$Fe^{3+} + PO_4^{3-} \leftrightarrow FePO_4 \downarrow \quad (5.4)$$

Again, it is almost unavoidable that $Fe(OH)_3$ will form and precipitate out of the wastewater stream if the pH value is alkaline.

5.6 Membrane processes for advanced wastewater treatment

An overview of membrane processes that can be used in wastewater treatment has been presented in Chapter 3; the general characteristics of various processes determine the applications of membrane processes in wastewater treatment. Membrane filtration can, theoretically, replace conventional processes such as secondary sedimentation, flocculation, settling basin, and granular filtration altogether. In reality, however, the applications of membrane filtration in wastewater treatment need to be carefully planned and implemented.

Membrane filtration is usually placed after the secondary treatment, when the wastewater has rid itself of the majority of suspended particulates and FOG. Cartridge filters or carbon filters are often used before the membrane unit for extending the working life cycle of the membrane material, which is very susceptible to fouling or to forming an adsorbed layer by lipids, proteins, silicates, and other minuscule substances.

Low-strength wastewater from food and agricultural processing can be treated with membrane filtration alone, provided that some forms of pre-treatment (filtration) precede the membrane filtration. For example: steep or soaking water from grain processing can be treated with a microfiltration unit; wastewaters from milk and cheese processing including cleaning water and evaporator condensate may be filtered with ultrafiltration or a combination of microfiltration and ultrafiltration; and oil/water emulation may be separated with a ceramic membrane filtration unit. A number of membrane filtration processes that are capable of retaining or removing certain materials are classified based on their size exclusion capability, as shown in Figure 5.4.

Reverse osmosis (RO) is often associated with water treatment or ultrapure water production because of its ability to retain the dissolved ions (in the case of ultrapure water production, RO serves as pre-treatment for ion-exchange deionization). One successful commercial application of reverse osmosis is desalination of seawater or brackish water and production of bottled water. RO is also

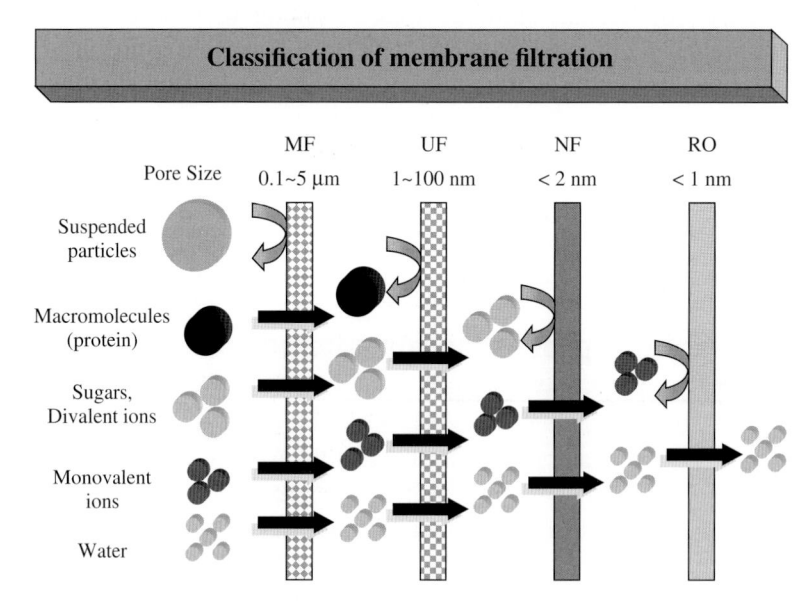

Figure 5.4 Classification of membrane filtration based on size exclusion (see plate section for color version)

used with some success in removing arsenic from drinking water sources, and it may also be used to desalt effluents from a wastewater treatment plant to reduce the salt concentration before discharge.

RO is widely used in food processing as a concentration process and recovery process of useful components in food wastewaters. It is conceivable that RO can also be used to reduce the volume of the wastewater in food wastewater treatment (e.g., it may be used to treat nutrients such as nitrate and phosphate), or to produce recycled water for reuse in food processing operations.

Like RO, electrodialysis (ED) is used to desalt impaired waters or remove ions such as nitrate, arsenic, and phosphate. ED may also be used to separate acids from food wastewaters. Recovery of carboxylic acids (e.g., acetic, citric, and lactic) is a known application of ED in food and agricultural processing.

As explained in Chapter 3, there are four common types of membrane module designs available for membrane processes. However, not all module types are suitable for all membrane processes. Table 5.1 provides a guideline for selecting modules for membrane processes.

One current interest in applications of membrane processes in food wastewater treatment is the recovery of valuable commodities from food wastewater streams. The entirety of Chapter 8 is devoted to the recovery of useful materials and energy from food and agricultural wastewaters, including using membrane-based technologies to achieve the objectives of the recovery.

One problem that has hindered the widespread use of membrane technology is the noticeable occurrence of concentration polarization/fouling in membrane processes. The detrimental effect of concentration polarization and/or membrane fouling adds significant costs to the operator. In membrane filtration processes, concentration polarization is formed as the result of the rapid accumulation of retained solutes near the membrane surface, to the point where the concentration of macromolecule solute reaches the gel-forming concentration and the retained molecules diffuses back into the bulk fluid. The cause of concentration polarization in pervaporation or electrodialysis is slightly different from that of

Table 5.1 Membrane modules for common membrane processes

Process	Tubular	Hollow fiber	Plate & frame	Spiral-wound
Microfiltration	Good	Not suitable	Good	Not suitable
Ultrafiltration	Good	Adequate	Good	Adequate
Nanofiltration	Good	Good	Good	Adequate
Reverse osmosis	Adequate	Good	Adequate	Good
Pervaporation	Adequate	Good	Good	Good
Electrodialysis	Not suitable	Not suitable	Good	Not suitable

membrane filtration, in that it is triggered by the relatively slow diffusional mass transfer rates of solutes or ions from the bulk to the membrane surface.

Membrane fouling is commonly observed as the membrane flux is continuously declining after a period of time of operation. This is distinguished from concentration polarization in being usually an irreversible, partially concentration-dependent and time-dependent phenomenon. The identification of membrane fouling often relies on the operator's experience, performing fouling tests with lab-scale static filtration experiments or silt density index (SDI) measurements, and membrane vendor's recommendations. Membrane fouling is intimately related to concentration polarization, but the two are not exactly interchangeable in our description of membrane performance deterioration. The exact cause of membrane fouling is very complex and, therefore, is difficult to depict in full confidence with available theoretical understandings. Fouling is influenced by a number of chemical and physical parameters, such as concentration, size of particulates, pore size distribution, temperature, pH, ionic strength, and specific interactions (hydrophobic interaction, hydrogen bonding, dipole-dipole interactions).

Membrane fouling can be greatly reduced in several ways. One effective way is to provide pre-treatment to the feed liquids. Some simple adjustments, such as varying pH values and using hydrophilic membrane materials, can also provide some relief from membrane fouling. There is also widespread interest around the world in modifying membrane properties to minimize the membrane-fouling tendency.

Since membrane fouling is intimately associated with concentration polarization phenomenon, any action taken to minimize concentration polarization will also help reduce membrane fouling. Fouled membranes can be cleaned and will regain some of their original performance, but frequent cleaning and washing with detergents will inevitably lead to the demise of the membrane. There are three basic types of cleaning methods currently used: hydraulic flushing (back-flushing), mechanical cleaning (only in tubular systems) with sponge balls, and chemical washing. When using chemicals to perform de-fouling, cautions must be observed, since many polymeric membrane materials are susceptible to chlorine, high pH solutions, organic solvents, and other chemicals.

5.7 VOC removal with pervaporation

As described in Chapter 3, pervaporation is an energy-efficient technology that has been used commercially for alcohol dehydration, VOC removal from contaminated water and hydrocarbon separations. The driving force of pervaporation processes is the chemical potential difference across the membrane between the feed and permeate; unlike distillation, the performance of pervaporation is not restricted by vapor-liquid equilibrium (Dutta et al., 1996). Recently, it has been shown to be a valuable tool for value-added wastewater treatment through flavor and aroma recovery from food processing by-products (e.g., Karlsson and Trägårdh, 1996; Peng et al., 2003c). The application of pervaporation in VOC removal has also been intensively researched (e.g., Jiang et al., 1997; Hitchens et al., 2001; Vane et al., 1999, 2001a; Peng and Liu, 2003a, 2003b; Liu and Peng 2006). These VOC removal research programs have led to several successful field demonstrations (e.g., Alvarez et al., 2001; Vane et al., 2001b).

In general, pervaporation processes can be easily adapted to VOC removal in water or wastewater because of their energy efficiency and targeted removal without introducing additional chemicals or new pollutants in different forms (e.g., carbon adsorption and air stripping). Almost all VOCs can be removed with pervaporation. However, VOCs of particular interest, including petroleum-based solvents such as benzene, toluene, ethyl benzene and xylenes (BTEX), and chlorinated solvents such as trichloroethylene (TCE) and tetrachloroethylene (PCE), are particularly well suited for pervaporation removal. The water solubilities of these compounds are low, so the amount of VOCs dissolved in water is too small to be economically removed from water by conventional chemical process separation technologies such as distillation. In the past, air stripping and/or activated carbon treatments were deployed for the task. However, the former is susceptible to fouling and merely turns a water pollution problem into an air pollution issue, while the latter needs costly regeneration steps and may not be suitable for VOCs that are easily displaced by other organic compounds. Over the decades, a growing literature has been added to the knowledge base of VOC removal with pervaporation.

5.8 Disinfection

Disinfection is a process in which pathogenic organisms are destroyed or inactivated by physicochemical treatments. It is a final step for water treatment but, increasingly, wastewater treatment plants also apply disinfection in wastewater treatment because of concern about pathogens in treated wastewater being discharged to a receiving water body or field. Usually, pathogens in raw wastewater will have been, to a large extent, either removed or inactivated along the trail of wastewater treatment processes. Nevertheless, there are still many opportunities for re-contaminating treated wastewater, since many treatment processes are conducted in open facilities outdoors. Other reasons for wastewater disinfection, or using disinfectants such as chlorine or its derivatives in wastewater treatment, are: oxidation of ammonia and of organic materials that contribute to BOD; destruction and control of ion-fixing and slime-forming bacteria; and destruction and control of filter flies, algae, and slime growth on trickling filters.

The mechanism of disinfection is said to be inactivation of enzymes in the pathogens by denaturing them; without functional enzymes, microorganisms are destroyed or inactivated. In order to gain access to the enzymes in pathogen cells, the cell walls have to be penetrated by disinfectants or destroyed by thermal, chemical or physical means. Chemical disinfectants, such as ozone, chlorines, and chlorine dioxide, work through oxidizing and reacting with cells of pathogens. Other techniques such as heat energy, irradiation, ultrasound and high pressure perform their duties through physical destruction.

Chlorine and its derivatives are the most common disinfectants used in water and wastewater treatment. Chlorine in aqueous solution hydrolyzes to yield

$$Cl_2 + 2H_2O \xrightleftharpoons{\text{reversible}} H_3O^+ + Cl^- + HOCl \qquad (5.5)$$

HOCl may be further hydrolyzed to yield:

$$HOCl + H_2O \xrightleftharpoons{\text{reversible}} H_3O^+ + OCl^- \qquad (5.6)$$

Both HOCl and OCl$^-$ are disinfectants. Although chlorine is a very effective disinfectant, handling it can be inconvenient, to say the least. Sodium hypochlorite (NaOCl) is often used in place of chlorine for disinfection of water and wastewater. Chlorinated lime (containing up to 70% CaOCl$_2$ and 20% Ca(OH)$_2$, as well as carbonate), also called bleaching powder, is also widely used the same manner as NaOCl in applications. Calcium hypochlorite (Ca(OCl)$_2$) is another chlorine derivative that is broadly employed in water and wastewater disinfection. Chlorine dioxide, an unstable gas, is also used in disinfection, and is often generated onsite through the following reaction:

$$2NaClO_2 + Cl_2 \xrightleftharpoons{\text{reversible}} 2ClO_2 + 2NaCl \qquad (5.7)$$

Another group of disinfectants comprise strong oxidizing agents. Ozone (O$_3$) is a particularly powerful but unstable oxidizing agent and is used extensively in Europe, both for disinfection and for removal of objectionable odor, state, and color. The popularity of ozone in Europe is also linked to its lack of residual products or byproducts that might be harmful to human health. However, ozone is unstable, and ozone-treated water does not have residual protection from re-contamination, as that treated with chlorine does. As a result, in the United States, chlorine and its derivatives are still dominant in water and wastewater disinfection. Hydrogen peroxide (H$_2$O$_2$) is an unstable liquid oxidizing agent, but its oxidizing power is somehow not entirely related to its disinfecting power. It is believed that some bacteria can produce an enzyme called catalase, that decomposes hydrogen peroxide into water and gaseous oxygen, thus rendering hydrogen peroxide harmless to those bacteria. Consequently, hydrogen peroxide is not suitable for disinfectant for any large-scale water or wastewater disinfection.

The disinfection or inactivation of microorganisms is, in a way, a physicochemical process that is not instantaneous. The rate of disinfection is believed to follow a first-order relationship called Chick's law:

$$-\frac{dN°}{dt} = kN° \qquad (5.8)$$

which can be integrated to result in

$$N° = N°_0 e^{-kt} \qquad (5.9)$$

where:
$N°$ is the number concentration of surviving microorganisms at time t
k is the rate constant.

Chick's law should be used as a rough estimation of rate of disinfection in a practical application of disinfectants. This rate may be increased or decreased, depending on the environmental factors, disinfectants, and microorganisms. Also, the concentration of disinfectants or dosage of disinfectants is not reflected in Equation (5.9), while temperature effect of rate of disinfection cannot be easily included in the rate constant, k, because temperature also affects certain reactions involved in disinfection, in addition to disinfection rate. The value of pH can also exert influence on disinfection rate as well as the reaction steps involved in certain disinfection processes. Extreme pH can inactivate microorganisms without disinfectants. Organic matter may interfere with disinfection processes by reacting with disinfectants or by shielding microorganisms that attach to the surfaces of organic matters.

Non-chemical disinfection may also be used in lieu of chemicals. Thermal treatment is an effective method of inactivation of microorganisms, and extended thermal treatment such as high-temperature steam or boiling water can achieve sterilization. However, thermal treatment is an unlikely choice of disinfection method for wastewater treatment, due to the enormous cost it entails. Ultraviolet irradiation has certain bactericidal effects, but its effectiveness is debatable, as the presence of various substances – including water itself – will deplete the strength of ultraviolet irradiation. Gamma- and X-ray irradiation can inactivate certain species of bacteria but, once again, this technology is impractical in disinfecting wastewater, as economical feasibility for the treatment of large volumes of water dictates the selection of disinfection methods.

5.9 Further reading

Henze, M., Harremose, P., Jansen, J.L.C. & Arvin, E. (2001a). *Wastewater Treatment: Biological and Chemical Processes.* Third edition. Springer, Berlin, Germany.

5.10 References

Abou-Nemeh, I., Majumdar, S., Saraf, A., Sirkar, K.K., Vane, L.M., Alvarez, F.R. & Hitchens, L. (2001). Demonstration of pilot-scale pervaporation systems for volatile organic compound removal from a surfactant enhanced aquifer remediation fluid. II. hollow fiber membrane modules. *Environmental Progress* **20**, 64–73.

Ahn, J. Daidou, T., Tsuneda, S. & Hirata, A. (2002). Characterization of denitrifying phosphate-accumulating organisms cultivated under different electron conditions using polymerase chain reaction-denaturing gradient gel electrophoresis assay. *Water Research* **36**, 403–412.

Aiyuk, S., Amoako, J., Raskin, L., van Haandel, A. & Verstraete, W. (2004). Removal of carbon and nutrients from domestic wastewater using a low investment, integrated treatment concept. *Water Research* **38**, 3031–3042.

Alvarez, F.R., Vane, L.M. & Hitchens, L. (2001). Demonstration of pilot-scale pervaporation systems for volatile organic compound removal from a surfactant enhanced aquifer remediation fluid. I. spiral wound membrane modules. *Environmental Progress* **20**, 53–63.

Arun, V., Mino, T. & Matsuo, T. (1987). Biological mechanisms of acetate uptake mediated by carbohydrate consumption in excess phosphate removal systems. *Water Research* **22**, 565–570.

Bond, P., Hugenholtz, P., Keller, J. & Blackall, L. (1995). Bacterial community structures of phosphate removing and non-phosphate removing activated sludge from sequencing batch reactor. *Applied Environmental Microbiology* **61**, 1910–1916.

Dutta, B.K., Ji, W. & Sikdar, S.K. (1996). Pervaporation: Principles and Applications. *Separation and Purification Methods* **25**, 131–224.

Fuhs, G.W. & Chen, M. (1975). Microbiological basis of phosphate removal in the activated sludge process for the treatment of wastewater. *Microbiology and Ecology* **2**, 119–138.

Henze, M., Harremose, P., Jansen, J.L.C. & Arvin, E. (2001b). *Wastewater Treatment: Biological and Chemical Processes.* 430pp, 3rd edition. Springer, Berlin, Germany.

5.10 REFERENCES

Hitchens, L., Vane, L.M. & Alvarez, F.R. (2001). VOC removal from water and surfactant solutions by pervaporation: a pilot study. *Separation and Purification Technology* **24**, 67–84.

Jiang, J.-S., Vane, L.M. & Sikdar, S.K. (1997). Recovery of VOCs from surfactant solutions by pervaporation. *Journal of Membrane Science* **136**, 233–247.

Karlsson, H.O.E. & Trägårdh, G. (1996). Applications of pervaporation in food processing. *Trends in Food Science & Technology* **7**, 78–83.

Kuba, T., Smolders, G.J.F., van Loosdrecht, M.C.M. & Heijnen, J.J. (1993). Biological phosphorus removal from wastewater by anaerobic and anoxic sequencing batch reactor. *Water Science Technology* **27**, 241–252.

Kuba, T., Wachtmeister, A., van Loosdrecht, M.C.M. & Heijnen, J.J. (1994). Effect of nitrate on phosphorus release in biological phosphorus removal systems. *Water Science Technology* **30**(6), 263–269.

Liu, S.X. & Peng, M. (2006). Assessment of semi-empirical mass transfer correlations for pervaporation treatment of wastewater contaminated with chlorinated hydrocarbons. *JZUS* (in press).

Mino, T., van Loosdrecht, M.C.M., Heijnen, J.J. (1998). Microbiology and biochemistry of the enhanced biological phosphate removal process. *Water Research* **32**, 3193–3207.

Maurer, M., Gujer, W., Hany, R. & Bachmann, S. (1997). Intracellular carbon flow in phosphorus accumulating organisms in activated sludge systems. *Water Research* **31**, 907–917.

Mulkerrins, D., Dobson, A.D.W. & Colleran, E. (2004). Parameters affecting biological phosphate removal from wastewaters. *Environment International* **30**, 249–259.

Peng, M. & Liu, S.X. (2003a). VOC removal from contaminated groundwater through membrane pervaporation. part II: 1,1,1- trichloroethane – surfactant solution system. *Journal of Environmental Sciences* **15**, 821–827.

Peng, M. & Liu, S.X. (2003b). VOC removal from contaminated groundwater through membrane pervaporation. part I: water-1,1,1- trichloroethane system. *Journal of Environmental Sciences* **15**, 815–820.

Peng, M. & Liu, S.X. (2003c). Recovery of aroma compound from dilute model blueberry solution by pervaporation. *Journal of Food Science* **68**, 2706–2710, 2003c.

Peng, M., Vane, L.M. & Liu, S.X. (2003). Recent Advances in VOC Removal from Water by Pervaporation. *Journal of Hazardous Materials* **B98**, 69–90.

Smolders, G.J.F., van Loosdrecht, M.C.M. & Heijnen, J.J. (1994). A metabolic model for the biological phosphorus removal process. *Water Science and Technology* **31**, 461–470.

Vane, L.M., Alvarez, F.R. & Giroux, E.L. (1999). Reduction of concentration polarization in pervaporation using vibrating membrane module. *Journal of Membrane Science* **153**, 233–241.

Vane, L.M., Alvarez, F.R. & Mullins, B. (2001). Removal of methyl t-butyl ether (MTBE) from water by pervaporation: bench-scale and pilot-scale evaluations. *Environmental Science & Technology* **35**, 391–397.

Vane, L.M., Hitchens, L., Alvarez, F.R. & Giroux, E.L. (2001). Field demonstration of pervaporation for the separation of volatile organic compounds from a surfactant-based soil remediation fluid. *Journal of Hazardous Materials* **81**, 141–166.

Wentzel, M.C., Loiter, L.H., Loewenthal, G.A. & Marais, G.v.R. (1986). Metabolic behavior of *Acinetobacter* spp. in enhanced biological phosphorus removal – a biochemical model. *Water SA* **12**, 209–224.

Zeng, R.J., Yuan, Z. & Keller, J. (2003). Model-based analysis of anaerobic acetate uptake by a mixed culture of polyphosphate-accumulating and glycogen-accumulating organisms. *Biotechnology and Bioengineering* **83**(3), 293–302.

6
Natural systems for wastewater treatment

6.1 Introduction

Natural wastewater treatment refers to a category of technologies that specifically and substantially utilize natural methods to reduce contaminants from wastewaters in large open fields without the necessity of energy-intensive mechanical equipment operations for major treatment responses. Natural wastewater treatment systems have enjoyed a revival not only in developing countries, but also in the United States. The impetus for this renewed interest in natural systems is a combination of cost-consciousness and a new mindset of recycling and reuse of yesteryear's rejected ideas.

Large-scale land application of wastewater has been practiced for more than 150 years. Initially, municipal and industrial wastewaters were routinely discharged into rivers and lakes. However, as the population of major cities in Western Europe and America grew exponentially, due to urban migration from rural areas and natural growth of population as the Industrial Revolution accelerated, the practice of discharging wastewater into rivers or lakes became a public scourge and the source of epidemic outbreaks. In London during the mid-19th century, the river Thames was constantly filled with human feces and other unmentionables, and the city was covered by a pall of stinking air. The curtains of the Houses of Parliament were soaked with chloride of lime to ward off the overwhelming odor during parliamentary sessions. It was not until Sir Edwin Chadwick took up the question of sanitation in 1843 that this indignity finally came to an end. Chadwick advocated separation of different sewers by practicing the principle of "the rain to the river and the sewer to the soil." This is believed to have been the beginning of large-scale land applications of municipal and industrial wastewaters.

The benefits of this practice of sewer disposal were quickly realized with its fertilizing capability. By the turn of the 20th century, almost all wastewater

Food and Agricultural Wastewater Utilization and Treatment, Second Edition. Sean X. Liu.
© 2014 John Wiley & Sons, Ltd. Published 2014 by John Wiley & Sons, Ltd.

generated in Western European cities and the North American continent was applied to the land. Today, natural systems, including land applications, are still used for wastewater treatment and management in many parts of the nation, although many different regulations and ordinances are enforced in different jurisdictions.

The natural systems for wastewater treatment are different from the wastewater treatment technologies described in the previous chapters, in the sense that the natural components of the treatment systems accomplish the majority of the process objectives. This means that natural systems for wastewater management and treatment do not involve large-scale energy and materials inputs.

Physical and biological wastewater treatment processes are often complex operations requiring intensive energy input (for mechanical devices/equipment) or/and material input (e.g., flocculants and oxidants) even though these processes also utilize natural components (e.g., gravity for sedimentation and screening; microorganisms for BOD and nutrient removal). Natural systems for wastewater treatment also provide silent, odor-free, and robust treatment processes. They do, however, require larger swath of land than those of conventional and more energy intensive treatment processes. Overall, the natural systems for wastewater treatment and management are categorized based on different environmental conditions: aquatic, terrestrial, and wetland.

Constructed wetlands, aquacultural operations, and sand filters are generally the most successful methods of treating the treated wastewater effluent from stabilization lagoons. These systems have also been used with more traditional, engineered primary treatment technologies, such as Imhoff tanks, septic tanks, and primary clarifiers. Their main advantage is to provide additional treatment beyond secondary treatment, where required.

In recent years, there has been a revival of the use of natural systems for agricultural wastewaters from intensive animal farming. North Carolina and other southern states of the USA have renewed interests in employing aquatic plants to treat animal wastewaters that contain large amounts of compounds of nitrogen and phosphorus. One of the plants used in this region is duckweed, which looks like "oversized" algae floating on a pond or river. Duckweed is one of the smallest flowering plants in the world, and it can be used as food for fish and various species of water birds, including, of course, ducks. The other application of the high-strength wastewater, in growing biomass for value-added bio-based materials or energy, also attracts interest in the animal farming industries in the region.

6.2 Stabilization ponds

One of the ancient wastewater treatment technologies, the stabilization pond (also referred to as lagoons), has been used continuously as a method of sewage

Figure 6.1 A waste stabilization pond system in Mèza, France. It contains two small anaerobic ponds, four experimental facultative ponds, and one series pond system composed of a facultative pond and four maturation ponds (see plate section for color version)

disposal. In some cases, such ponds were also utilized for aquaculture. Stabilization ponds are used for both municipal wastewater treatment and industrial wastewater treatment, particularly for wastewaters from small communities and seasonal industrial wastewaters, as well as for less affluent communities throughout the world (Figure 6.1).

Although stabilization ponds can be used in most regions of human inhabitation, their best performances in treating wastes are in warm climates with adequate sunlight. The current interest in waste stabilization ponds (WSPs) is a result of the accidental discovery of their capabilities when WSPs were used initially as simple sedimentation basins or emergence holding ponds at wastewater treatment plants. A WSP is a relatively shallow body of wastewater contained in an earthen man-made basin, into which wastewater flows, and from which, after a certain retention time (time that takes the effluent to flow from the inlet to the outlet), a well-treated effluent is discharged.

Many characteristics make WSP substantially different from other wastewater treatment, including design, construction and operation simplicity, cost effectiveness, low maintenance requirements, low energy requirements, easily adaptive for upgrading, and high efficiency. They are used for sewage treatment in temperate and tropical climates, and provide one of the simplest, lowest-cost and most efficient wastewater treatment technologies available. Waste stabilization ponds are very effective in the removal of fecal coliform bacteria.

Solar energy is the only requirement for the operation of a WSP. They have been in use in the United States since 1901 but, until fairly recently, the United States Environmental Protection Agency (US EPA) has formalized the design and operational criteria for WSPs (US EPA, 1983).

Over 7,500 WSP systems have been built in the United States, serving rural communities with a mean population less than 10,000. In Canada, there are over 1,000 WSPs in operation, representing about a half of the wastewater treatment capacity in the country (Townshend & Knoll, 1987). WSPs are the most common type of wastewater treatment systems in many other countries. This is particularly so in warmer climates (e.g., the Middle East, Africa, South Asia and Latin America), where ponds are commonly used for large populations (up to around 1 million). In developing countries, and especially in the tropical and equatorial regions, sewage treatment by WSPs has been considered an ideal way of using natural processes to improve sewage effluents.

In recent decades, Europe has seen an increased number of WSPs built in many parts of the continent where small populations reside (Gomes de Sousa, 1987). In France, there are approximately 2,500 WSPs in operation throughout the country, mostly for small communities (Racault *et al.*, 1995). The capital and operational costs of WSPs are lower than other comparable wastewater treatment technologies (Mara & Pearson, 1998).

Waste stabilization ponds can be used either alone or in combination with other wastewater treatment processes. A typical system consists of several constructed ponds operating in series, and treatment of the wastewater occurs as constituents are removed by sedimentation or transformed by biological and chemical processes. In the bottom of the ponds, a sludge layer is formed, which can impact performance by changing the pond's hydraulics due to a decrease in the pond's effective volume and also by changing the shape of the bottom surface. Therefore, periodic sludge removal is usually required. Stabilization ponds suitable for wastewater treatment are those that maintain right biological conditions for biological interactions and reactions that break down the organic matters and inorganic nutrients.

There are four basic types of WSPs. All use microorganisms to degrade and decontaminate organic and inorganic constituents. The types of organisms, as well as the amount of oxygen present in the pond systems, differ among the five categories forming the basis for classifying stabilization ponds:

- Facultative ponds
- Maturation ponds
- Aerated ponds
- Aerobic ponds
- Anaerobic ponds

The term "oxidation pond" is also used to describe any pond system that utilizes oxygen to break down organic matters, either with or without a significant portion of the dissolved oxygen (DO) provided by photosynthetic algae. All stabilization ponds, with the exception of anaerobic ponds, can be grouped into the oxidation pond category. Recently, one type of oxidation pond, called High Rate Algae Pond (HRAP), gained increasing attention from both wastewater treatment researchers and bioenergy experts due to its application in biofuel production. Biofuels are generated from microalgae biomass produced in this type of pond, where agricultural and food wastewater or even municipal wastewater are treated (see Chapter 8 for information on biofuel production in HRAPs).

6.2.1 Facultative ponds

Facultative ponds are the commonest type of stabilization ponds in use, being able to treat completely both raw, settled sewage and also a wide range of industrial wastewaters, including food and agricultural wastewaters, with detention time of 5–30 days. These ponds have depths ranging from 1.2–1.5 m (4–8 ft), consisting of two layers of biological treatment zones – an aerobic layer on top of an anaerobic layer, often containing sludge. The top aerobic layer stabilizes the wastewater, while fermentation takes place in the anaerobic bottom layer. The oxygen needed for aerobic stabilization comes from photosynthesis of algae in the pond (Figure 6.2). This oxygen is used by heterotrophic bacteria in the pond for the aerobic breakdown of organic matter, and respiration of organic matter provides a source of carbon dioxide for the algae. This symbiotic relationship between the algae and the bacteria provides the basis for this natural method of wastewater treatment.

Since algae in the discharged effluent needs to be removed, the treatment process usually involves a facultative pond, one or two maturation ponds and a tertiary treatment phase such as rock filters or intermittent sand filtration for the removal of algae before discharge of the final effluent to an adjacent watercourse. Total containment ponds are suitable for climates where evaporative loss of water exceeds the rainfall; controlled discharge ponds are used in many areas and climates. The controlled discharge ponds operate discharges one or twice a year, depending on the quality of the treated water being discharged, and tend to have longer detention times. The longer detention time, and the issue of blooming algae in the treated water, offset the savings associated with facultative ponds.

An obvious solution to the inherited drawbacks of facultative ponds is to add oxygen into the ponds. In an aerated pond, oxygen is supplied through mechanical aeration equipment and air diffusers. These ponds (which could still be classified either as facultative or as aerobic, depending on their DO profile), unlike facultative ponds, can be built deeper, usually in the range of 2–6 m (6–20 ft), because oxygen can be introduced at greater depths. As a result, the detention

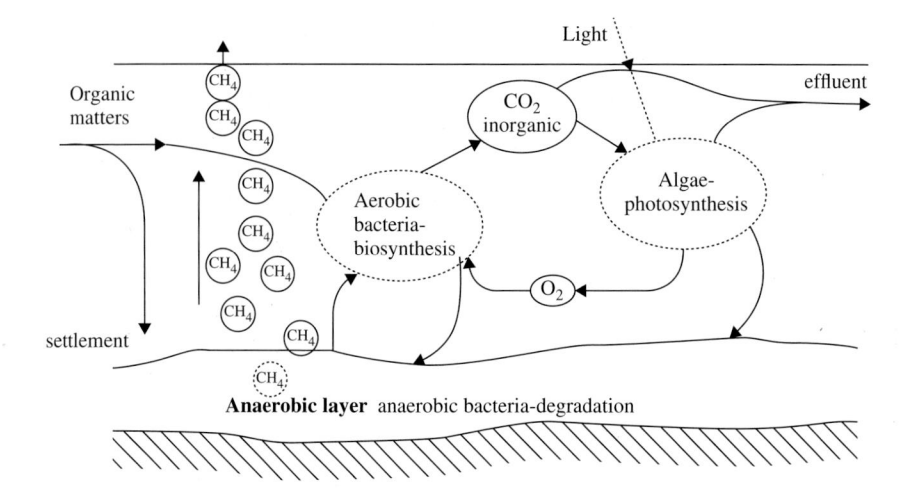

Figure 6.2 A schematic diagram of a facultative pond with description of interactions among its components

time is shortened and is in the neighborhood of 3–10 days. Depending on the method used to deliver oxygen into the pond, an aerated pond can be designed as completely-mixed or partially-mixed reactor, thus reducing the footprint of the pond. A completely-mixed reactor used for an aerated pond has many design characteristics similar to an activated sludge reactor.

6.2.2 Maturation ponds

Maturation ponds, sometimes also called tertiary-maturation ponds, are low-rate stabilization ponds designed to provide secondary effluent (conventional secondary processes or facultative ponds) polishing and pathogen removal. The mechanism for pathogen removal in maturation ponds is actually simple: the removal of microorganisms (pathogens) is due to natural die-off, predation, sedimentation, and adsorption; the majority of pathogens settle onto the sludge in the bottom of the ponds, thus removed from effluents. In fact, all pond systems can, to some degree, remove pathogens from wastewaters. In order for maturation ponds to remove pathogens substantially, however, detention times in these ponds must be long enough for pathogens to settle down in the ponds. There is some risk if the sludge is removed from the maturation ponds, so handling of the sludge requires caution and it needs to be treated or kept away from public access.

For high-strength wastewaters, maturation ponds are used for improving the effluent quality prior to surface water discharge of treated wastewaters. Their design, size, and number, in series, are decided in many parts of the world based on the need of removing pathogens from treated wastewater. If the objective of using maturation ponds is for maximum purity in terms of BOD_5 reduction and pathogen removal, algae production is discouraged in order to achieve maximum light penetration.

Maturation ponds, however, are of the same depth as facultative ponds (1–1.5 m or 3–4.5 ft). A minimum of 15–20 days is used as detention time for maturation ponds. Nitrogen removal is achieved in these ponds through denitrification in the settled sludge. Phosphorus may also be removed by a diverse range of algal communities. However, algal growth is not desirable, especially if the effluent is slated to discharge into the receiving water, because of turbidity and suspended solids caused by algae. Rock filters in submerged beds are sometimes used for removal of these algal solids. The algae retained by rock filters decompose and are utilized by bacterial biofilm on rock materials. In general, the arrangement of pond systems starts with an anaerobic pond, then a facultative pond, and finally a maturation pond in series, depending on the quality of the influent and the treatment objective.

6.2.3 Aerated lagoons

An aerated lagoon is a stabilization pond with its aerobic condition maintained by mechanical or diffused aeration equipment. Bubble aeration is sometimes provided to keep the pond aerobic in locations where pond surfaces are frozen for extended periods in winter. Unlike aerobic ponds, aerated lagoons do not rely on algae for oxygen delivery. The microbial characteristics of an aerated pond are very similar to those of an activated sludge process. Its detention times are in the order of 1–10 days, depending on organic loading rate, temperature, and the degree of treatment required. The organic loading of aerated lagoons is expressed as BOD_5 per unit volume per day. Aeration ponds are susceptible to large amount of BOD_5 loadings or toxic wastes, which can severely hinder the ponds' efficiency. The solids in the pond need to be suspended all the time in order to avoid the solids settling in the bottom of the pond and forming an anaerobic layer, thus reducing the efficiency and generating odor. Figure 6.3 shows a schematic depiction of an aerated lagoon with an aerator at work.

6.2.4 Aerobic ponds

Aerobic ponds, also called high-rate aerobic ponds, are designed to maintain a constant dissolved oxygen (DO) level throughout their depth. They can be

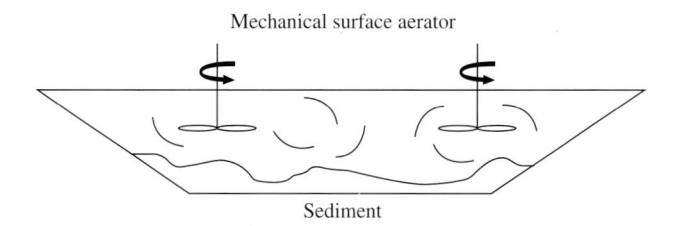

Figure 6.3 A schematic diagram of an aerated pond with surface aerators

viewed as "turbo-charged" facultative ponds with added oxygen functioning. As a unit process, aerobic ponds fall between facultative ponds and activated sludge processes. They are usually shallow (30–45 cm or 12–18 in), allowing light to penetrate their entire depth. Oxygen is provided by both an external device and algae through photosynthesis; mixing is also provided, to disperse oxygen and expose algae to sunlight, as well as to prevent anaerobic conditions in the ponds. The detention time for this type of ponds is relatively short, for about 3–5 days. In order to maintain the constant oxygen level in all depths, these ponds are best used in warm and sunny locations, where photosynthesis of algae is quick enough to provide sufficient DO, which is needed to degrade the organic matters in wastewater.

6.2.5 Anaerobic ponds

For strong food and agricultural wastewater, anaerobic ponds can be used to degrade the heavy loading of organic matters. In general, there are three identifiable zones in a basic design of an anaerobic lagoon: the scum layer, the supernatant layer, and the sludge layer. The system requires longer time (20–50 days), its depth is in the range of 2.5–5 m (8–15 ft), and it produces acid and methane. In anaerobic ponds, BOD_5 removal is achieved by sedimentation of solids, and subsequent anaerobic digestion in the resulting sludge. The process of anaerobic digestion is more intense at temperatures above 15°C. The anaerobic bacteria are usually sensitive to pHs lower than 6.2, so acidic wastewater must be neutralized prior to its treatment in anaerobic ponds.

A well-designed anaerobic pond will achieve about 40% removal of BOD_5 at 10°C, and more than 60% at 20°C. A shorter retention time of 1.0–1.5 days is commonly used. The design criteria for anaerobic lagoons are different from other designs of the stabilization ponds, the main noticeable difference being the depth of the lagoon. Oxygen transfer through the air-water interface of a lagoon is not important at all; in fact, it is undesirable for the anaerobic lagoon. Thus, anaerobic lagoons are typically deep basins with 2–5 meters (6.5–16.3 ft) in depth.

Figure 6.4 Photograph of an anaerobic pond for wastewater treatment in the field (see plate section for color version)

Anaerobic lagoons are often used as a preliminary treatment for high-strength wastewaters with a high content of organic materials, such as those found in food processing wastewaters rich in fat and proteins. As a result, anaerobic lagoons can only partially stabilize wastewaters, and further treatment – most likely aerobic processes – is needed before the wastewaters can be discharged to the receiving waters. It is common to place facultative ponds or activated sludge processes right after anaerobic lagoons for desired treatment goals. Slaughterhouses and meat processing wastewaters (BOD: $\approx$1,400 mg/L; FOG: $\approx$500 mg/L; pH: $\approx$7; temperature: 28°C) are particularly suitable for treatment by anaerobic lagoons after preliminary treatment such as removal of coarse solids and excessive grease or blood. Figure 6.4 shows a picture of an anaerobic stabilization pond in the field.

The production of methane in anaerobic lagoons (or any other anaerobic process) could be a potentially troubling issue in the future, as methane is a far worse greenhouse gas than carbon dioxide.

6.2.6 Design of stabilization ponds

WSPs often comprise a single string of anaerobic, facultative or aerated/aerobic, and maturation ponds in series, or several such series in parallel, depending on the organic strength of influents and the effluent quality objectives. In essence, anaerobic and facultative or aerated/aerobic ponds are designed for removal of BOD, and maturation ponds for pathogen removal. For ease of maintenance and flexibility of operation, at least two trains of similar ponds in parallel are built into many designs.

High-strength wastewaters, with BOD_5 concentration in excess of about 300 mg/L, will frequently be introduced into first-stage anaerobic ponds, which achieve a high volumetric rate of removal. However, wherever anaerobic ponds are deemed unacceptable by the regulatory agencies or the public at large, wastewaters, regardless of strength, may be discharged directly into primary facultative ponds.

Effluent from first-stage anaerobic ponds will overflow into secondary facultative ponds, which comprise the second stage of biological treatment. Pescod & Mara (1988) summarized several common pond system configurations (Figure 6.5), though other combinations may conceivably be used. The design loadings for stabilization ponds are usually measured in BOD per unit area. In anaerobic ponds, the loading is expressed as BOD per unit volume, because light, which is strongly affected by the surface area, is not an important factor in these ponds.

Facultative pond design is based on BOD removal, and most states set up their own design criteria for BOD loading and/or hydraulic detention time for facultative ponds. Based on their experience, Reed *et al.* (1995) recommended the loading rates shown in Table 6.1 for various climatic conditions for designing facultative ponds. Over the years, several empirical and rational models have

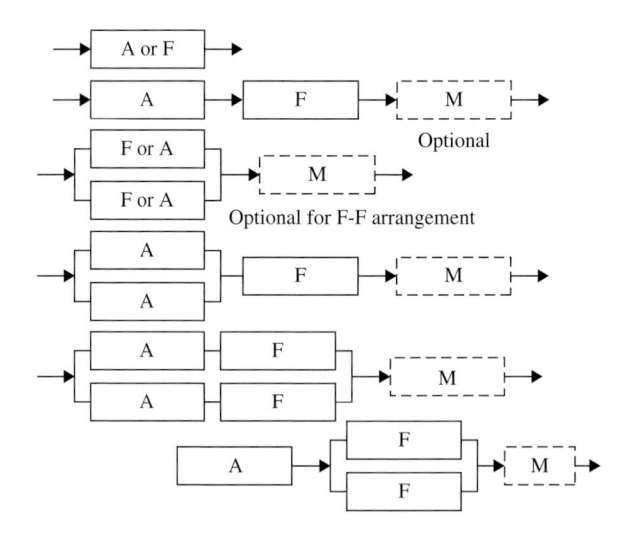

Figure 6.5 Several stabilization pond configurations: A = anaerobic pond; F = facultative pond; M = maturation pond. *Source*: Pescod & Mara (1988). Reproduced with permission of Elsevier

6.2 STABILIZATION PONDS

Table 6.1 Recommendations for loading rates for various climatic conditions

Climatic condition	BOD$_5$ loading rate
Air temperature above 15°C in winter	45–90 kg/ha·d (40-80 lb/ac · d)
Air temperature between 0–15°C in winter	22–45 kg/ha·d (20-40 lb/ac · d)
Air temperature below 0°C in winter	11–22 kg/ha·d (10-20 lb/ac · d)

Source: Adapted from Reed *et al.* (1995).

been developed for the design of the ponds (Reed *et al.*, 1995). These range from the ideal plug flow model to the completely mixed pond model, as well as models between these two extremes. Each of these models has produced satisfactory results in certain circumstances, but their use in the real world is limited because they all require evaluating multiple coefficients, which have to be evaluated from the similar WSP systems. Reed *et al* (1995) have compared several design models for facultative ponds and have concluded that there is no "best procedure" for recommendation in designing facultative ponds.

The BOD loading rate in the first cell is limited to 40 kg/ha·d or less, and the total hydraulic detention time in the system is 120–180 days when the average temperature of below 0°C and the loading can be increased to 100 kg/ha·d when the average temperature is above 15°C (Reed *et al.*, 1995).

The size of facultative ponds

$$L_s = 10 L_i Q / A_f \qquad (6.1)$$

where:
L_s = Surface BOD$_5$ loading, kg/ha·d
L_i = influent BOD$_5$, mg/L
Q = flow, m^3/d
A_f = facultative pond area, m^2

There is no consensus when it comes to the best design approach for anaerobic ponds. Customarily, volumetric BOD$_5$ loading rate, temperature, and hydraulic detention time have formed the basis for anaerobic ponds. The World Health Organization recommends the following guidelines for BOD removal rate 50% or higher (WHO, 1987):

- Volumetric loading, L_v, up to 300 g BOD$_5$ kg/m^3·d
- Hydraulic detention time of about 5 days
- Depth of the pond between 2.5 and 5 m

Table 6.2 BOD$_5$ reduction as a function of detention time and temperature

Temperature (°C)	Detention time (d)	BOD$_5$ removal rate (%)
10	5	0–10
10–15	4–5	30–40
15–20	2–3	40–50
20–25	1–2	40–60
25–30	1–2	60–80

Source: WHO (1987). Reproduced with permission of WHO.

Table 6.2 shows BOD$_5$ reduction as a function of detention time and temperature (adapted from WHO, 1987).

The size of anaerobic ponds

Anaerobic ponds can be satisfactorily designed, and without risk of odor nuisance, on the basis of volumetric BOD$_5$ loading (l_v, g/m^3·d), which is given by:

$$l_v = L_i Q / V_a \qquad (6.2)$$

where:
L_v = volumetric BOD$_5$ loading, kg/m^3·d or lb/ft^3·d
L_i = influent BOD$_5$, mg/L
Q = flow, m^3/d or ft^3/d
V_a = anaerobic pond volume, m^3 or ft^3

Hydraulic balance

To maintain the liquid level in the ponds, the inflow must at least be greater than net evaporation and seepage at all times. Thus:

$$Q_i = 0.001 \, A \, (e + s)$$

where:
Q_i = inflow to first pond, m^3/d or ft^3/d
A = total area of pond series, m^2 or ft^2
e = net evaporation (i.e., evaporation less rainfall), mm/d or in/d
s = seepage, mm/d or in/d.

Many WSPs are designed to handled with BOD$_5$ and TSS loadings, and their capacity of removing BOD$_5$ and Total Soluble Solids (TSS) is well documented and reasonably representative in their designs. However, nitrogen removal capacity of these ponds is crucial, given the fact that food and agricultural wastewaters tend to be rich in nitrogen content and that ammonia in treated wastewaters can be detrimental to fishes in the receiving waters. In the case of sludge from pond systems used in land treatment, nitrogen in the sludge may affect the design land application and its costs. Two important empirical models have been developed, based on the real data from the fields, and these have been validated by the data in later studies. The two models are summarized in Tables 6.3 and 6.4.

Phosphorus removal in stabilization ponds is an important design consideration only in the north central USA and Canada. For example, the discharge limit for phosphorus in the Great Lakes is 1 mg/L. In order to achieve this criterion,

Table 6.3 Design model for nitrogen removal from stabilization pond systems

---

For temperatures below 20°C is:
$C_e = C_i / \{1 + [(A/Q)(0.0038 + 0.000134T)\exp((1.041 + 0.044T)(pH-6.6))]\}$
and for temperatures above 20°C:
$C_e = C_i / \{1 + [5.035 \times 10^{-3}(A/Q)][\exp(1.540 \times (pH-6.6))]\}$

Where:
C_e = ammonia-N concentration in pond effluent, mg N/L
C_i = ammonia-N concentration in pond influent, mg N/L
A = pond area, m^2
Q = influent flow rate, m^3/d
pH = 7.3 exp(0.0005 A$_i$)
A_i = influent alkalinity, mg CaCO$_3$/L
Source: Pano and Middlebrooks (1982). Reproduced with permission of the Water Environment Federation.

Table 6.4 Design model for nitrogen removal from stabilization pond systems

---

$C_e = C_i \exp\{-[0.0064 (1.039)^{T-20}][q + 60.6 (pH-6.6)]\}$

Where:
C_e = total nitrogen concentration in pond effluent, mg N/L
C_i = total nitrogen concentration in pond influent, mg N/L
T = temperature, °C (range: 1–28°C)
q = retention time, d (range 5–231 d)
pH = 7.3 exp(0.0005 Ai)
A_i = influent alkalinity, mg CaCO$_3$/L
Source: Reed *et al.* (1995).

chemicals are added to the pond system to reduce the phosphorus level. A typical alum dosage of 150 mg/L may reduce phosphorus to less than 1 mg/L, and BOD$_5$ and TSS to less than 20 mg/L (Reed *et al.*, 1995).

Without chemical addition, phosphorus removal in stabilization ponds can still be achieved. The mechanisms of phosphorus removal most likely take place in maturation ponds (Mara *et al.*, 1992). The efficiency of total phosphorus removal in WSPs depends on how much biomass has settled in the bottom of the pond systems, as the biomass utilizes the phosphorus in wastewater (as described in the section on biological phosphorus removal in Chapter 5). Thus, the best way of increasing phosphorus removal in WSPs is to increase the number of maturation ponds so that, little by little, phosphorus becomes immobilized in the sediments. From a well-running two-pond system, 70% mass removal of total phosphorus may be achievable.

6.3 Land treatment systems

Land application of wastewater is perhaps the oldest method for disposal and treatment of wastewater. Early systems were used in England as "land farms" that received untreated wastewater and night soil from nearby communities. Today, land application systems have included application to edible and non-edible crops, to rangelands, to forests and wood plantations, to recreational areas including parks and golf courses, and to disturbed lands such as mine spoil sites.

Land application of wastewater also plays a role in recharging groundwater and recycling of fresh water. On average, water used once and then discharged to the ocean would not return as rain on land for about 2,600 years. This fact has drawn attention to the issue of wastewater reuse. Wastewater reuse in agriculture and other fields are not new, but increasing environmental awareness has made the reuse of wastewater, even after careful treatment, a tainted word. This level of concern is not unreasonable, given the checkered history of wastewater disposal throughout across the world. However, as wastewater treatment technologies advance and the quality of treated effluents steadily improves, the land application of treated wastewater from food and agriculture become a cost-effective alternative to discharging into the surface water, including the oceans.

There are two major categories of reuse of wastewater, which have been practiced throughout the world: potable use and non-potable use. The potable use of wastewater mainly includes returning reclaimed water to the drinking water supply after multiple levels of treatment, or using natural systems (including land applications) to treat wastewater directly. Non-potable uses of wastewater are many: the most promising examples include direct irrigation of fields using food wastewater with low BOD$_5$ and TSS; irrigation of parks, forests or golf courses with low-load wastewater; and use for aquaculture.

In many areas of the world, wastewater reuse has been practiced using a combination of treatment technologies that achieve a very high degree of treatment. Over the past 20 years, many states in the western USA have been treating wastewater to tertiary treatment standards and then allowing the wastewater to be reused, for irrigation or for recharge to groundwater aquifers. While this is an effective method for many arid regions in the western US, it is very expensive and is rarely practiced in other regions of the world, although there have been several studies conducted in Mexico and Egypt (Zachritz et al., 2001).

The land application of wastewater is not without risk of contamination of soil and groundwater beneath it. The challenge is to utilize the physicochemical and biological properties of the soil as an acceptor for the wastewater streams without any detrimental effect on the crops that are to be grown, to the ecosystem of which the land is part, to the characteristics of the soil, or to the quality of the groundwater and the surface water. The soil and the wastewater of a land application should be managed as an integrated system to obtain the best outcome.

Land application of wastewaters incorporates organic and inorganic materials into the soil for recycle and reuse. The assimilative capacity of a soil depends on its characteristics and the environmental conditions. The maximum capacity of a soil represents the maximum wastewater loading of the soil, and this is true for raw wastewaters as well as for treated wastewaters. Each application site will have a controlling parameter dependent on the characteristics of the wastewater applied, the characteristics of the soil and, most importantly, the environmental ramification. In most cases, permits are required for applying wastewater to the land.

Figure 6.6 is a schematic diagram of the use of land for plantation and wastewater land applications. As shown in the diagram, many factors are involved in the overall effect of water cycle on plants, including land application of wastewater. In most cases, the treated wastewater is applied to the land surface via furrow-flood, sprayer, or drip irrigation. BOD$_5$, TSS, and Fecal Coliform (FC) are partially removed in the conventional wastewater treatment steps, while the land application system removes additional BOD$_5$, TSS, and FC, as well as nitrogen and phosphorus.

Various designs of land application systems have been developed, including application of wastes to the soil surface using *slow rate*, *overland flow*, and *rapid infiltration* land treatment systems, and to the subsurface using leaching fields and absorption beds. The suitability of a particular system depends on the site characteristics, including: soil properties; ground topography such as slope and relief; local hydrology; groundwater depth and quality; land use; climatic factors such as temperature, precipitation, evapotranspiration, wind, and length of

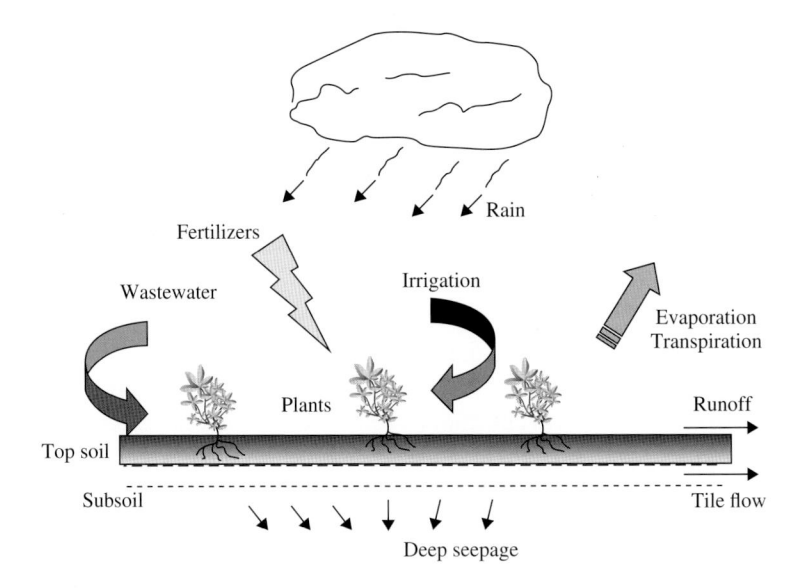

Figure 6.6 Schematic illustration of land applications of wastewaters

the growing season; and expected waste loading rates, as well as consideration of possible social and economic constraints. Tables 6.5 and 6.6 list the characteristics of three major types of land application systems and comparison of site characteristics for natural systems.

In designing any land application system, several common attributes of land application systems are often included in overall considerations; they are

- public health;
- groundwater issues;
- site evaluation and selection;
- crop selection and management if needed;
- pre-application treatment;
- distribution methods;
- design hydraulic loading rate;
- application rate;
- climatic consideration and storage;
- BOD_5 loading rate if needed;
- nitrogen removal;

- limiting nitrogen loading rate if required;
- suspended solid removal;
- phosphorus removal;
- land requirements;
- recovery of effluent from land application if possible.

6.3.1 Slow-rate (SR) systems

Slow rate systems are the prevailing form of land application system in use today. They involve the application of pretreated wastewater to vegetated soil to provide treatment of the wastewater and meet growth needs of the existing vegetation

Table 6.5 Characteristics of land wastewater treatment systems

Feature	Slow rate	Overland flow	Rapid infiltration
Application techniques	Sprinkler or surface[a]	Sprinkler or surface	Usually surface
Annual loading rate, m	0.5–6	3–20	6–125
Field area required, ha[b]	23–280	6.5–44	3–23
Typical weekly loading rate, cm	1.3–10	6–40[c]	10–240
Minimum pre–application provided in the USA	Primary sedimentation[d]	Grit removal and screening	Primary sedimentation[e]
Disposition of applied wastewater	Evapotranspiration and percolation	Mainly surface runoff and evapotranspiration	Percolation
Need for vegetation	Required	Required	Optional

[a] Includes ridge-and-furrow and border strip.
[b] Field area in hectares, not including buffer area, roads, or ditches for 3,785 m³/d (1 Mgal/d) flow.
[c] Range includes raw wastewater to secondary effluent, higher rates for higher level of pre-application treatment.
[d] With restricted public access; crops not for direct human consumption.
[e] With restricted public access.
Source: U.S. Environmental Protection Agency 1981, 1984.

Table 6.6 Comparison of site characteristics for land treatment processes

Characteristics	Slow rate	Overland flow	Rapid infiltration
Slope	Less than 20% on cultivated land; less than 40% on non-cultivated land	Finish slopes 1–8%[a]	Not critical; excessive grades require much earthwork
Soil permeability	Moderately slow and moderately rapid	Slow (for cold clays, silts, soils with impermeable barriers)	Rapid (sands, sandy loams)
Depth to groundwater	0.6–1 m (minimum)[b]	Not critical[c]	1 m during flood cycle[b]; 1.5–3 m during drying cycle
Climatic restrictions	Storage often needed for cold weather heavy and during heavy precipitation	Storage needed for cold weather	None (possibly modified storage needed in cold weather)

[a]Steeper grades might be feasible at reduced hydraulic loadings.
[b]Under-drains can be used to maintain this level at sites with high groundwater table.
[c]Impact on groundwater should be considered for more permeable soils.
Source: U.S. Environmental Protection Agency 1981, 1984.

through evapotranspiration and percolation (see Figure 6.7 (a)(b)(c)). The systems are similar to those found in common agricultural irrigation. The annual loading rate of the systems ranges from 0.5 to 6 m/year. There are roughly three types of slow-rate systems, categorized by the objectives of their applications (Paranychianakis *et al.*, 2006):

- Type I is based on the objective of reusing wastewater for crop and vegetation growth. The application of pretreated wastewater can be achieved through the uses of sprinklers, ridge-and-furrow, border strip flooding, and other surface or subsurface methods.

- Type II focuses on treatment of wastewater, not on reuse of wastewater. When the primary objective of the SR process is treatment, the hydraulic loading is usually limited either by the hydraulic capacity of the soil or the nitrogen removal capacity of the soil-vegetation matrix. Under-drains are sometimes needed for development of sites with high groundwater tables, or where perched water tables or impermeable layers prevent deep percolation. Perennial grasses are often chosen for the vegetation, because of their high nitrogen uptake, a longer wastewater application season, and the avoidance of annual planting and cultivation. Corn and other crops with higher market values are also grown on systems where treatment is the major objective.

- Filter beds treat wastewaters through the actions of percolation and filtration of the beds. The treated effluents are collected by a vast underground drain system. Filter bed systems are effective in nutrient removal and are suitable for population-intensive areas such as urban areas. Specifically modified filter beds can be used for treating various industrial wastewaters, such as those from animal farming and dairy operations (Jayawardane *et al.*, 1997).

SR system design

Slow-rate systems must be combined with other processes in order to produce a complete wastewater treatment. Pre-application treatment is often required to

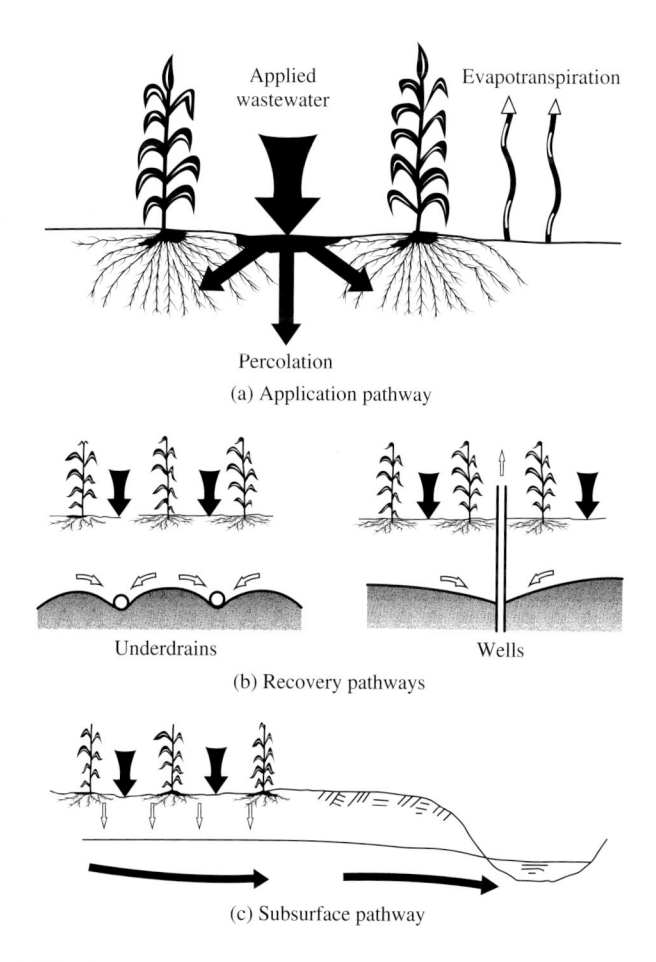

Figure 6.7 (a)(b)(c) A schematic diagram of a slow-rate wastewater treatment system. *Source*: U.S. Environmental Protection Agency 1981

ensure protection of public health, nuisance control, and distribution systems constraints. The main concern is the pathogen content in wastewaters. There are several common processes that can be used: fine screening, primary sedimentation and pH adjustment, as well as biological processes such as stabilization ponds.

Crop selection is also important for slow-rate land systems for application of pretreated wastewaters, particularly for water reuse. The choice of individual crop depends on the nitrogen content and revenue for Type I systems, and on maximum nutrient uptake for Type II. Revenue-generating field crops, such as corn, soybean, sugar beets, barley and wheat, are all good choices. For forage crops, alfalfa, quack grass ryegrass, orchard grass, bromegrass, and Kentucky bluegrass are suitable crops for maximum removal of nutrients.

For hydraulic loading for Type I system, the water balance equation is used to determine the hydraulic rate:

$$L_w = ET - P_r + P_w \tag{6.3}$$

where:
L_w = wastewater hydraulic loading rate based on soil permeability
ET = evapotranspiration rate, cm/month or in/month
P_r = precipitation rate, cm/month or in/month
P_w = percolation, cm/month or in/month

The ET is normally the monthly average ET rate of the selected crop, and it is determined from the historical evaporation data (at least 15 consecutive years). The value of P_r should be determined form a frequency analysis of wetter than normal years (using ten years as a basis). The design percolation rate P_w is estimated to be 4–10% of the measured field test data or published data based on wet/dry ratio, thus:

$$P_w(daily) = K\ (24\ h/d)(Adjustment\ Factor) \tag{6.4}$$

where:
P_w = design percolation rate, cm/d or in/d
K = permeability of limiting soil layer, cm/h or in/h
Adjustment Factor = 4–10% to account for wet/dry ratio and ensure a conservative and safe value for infiltration of wastewaters.

For hydraulic loading based on nitrogen limits, the nitrogen balance for the SR system is:

$$L_n = U + f(L_n) + A(C_p)(P_w) \tag{6.5}$$

6.3 LAND TREATMENT SYSTEMS

where:

L_n = mass loading of nitrogen, kg/ha·yr or lb/ac·yr
U = crop uptake, kg/ha·yr or lb/ac·yr
F = fraction of applied nitrogen lost to denitrification, volatilization, and soil storage
A = unit conversion factor, 0.1 for metric units and 2.7 in imperial units
C_p = percolation nitrogen concentration, mg/L, usually set at 10 mg/L due to the limiting nitrogen concentration
P_w = percolation rate, cm/yr or ft/yr.

Crop uptake can be found from the literature, and f value is based on wastewater characteristics and climate: 0.5–0.8 if BOD_5-to-nitrogen ratio is 5 or more; 0.25–0.5 for primary treatment effluents in municipal wastewater plants; 0.15–0.25 should be used for effluents from secondary treatment processes of municipal wastewater treatment plants; and a value of 0.1 is used for effluents from advanced wastewater treatment processes.

For Type II slow-rate systems, hydraulic loading rate is based on water balance and is expressed as:

$$L_w = IR - P_r \qquad (6.6)$$

where:

L_w = hydraulic loading rate
IR = crop irrigation requirement
P_r = precipitation, cm/yr, m/yr or in/yr

The crop irrigation requirement depends on the crop ET, the irrigation efficiency, and the leaching requirement. Incorporating these three factors into Equation (6.6) yields:

$$L_w = (ET - P_r)(1 + LR)(100/E)$$

where:

LR = leaching requirement
P_r = precipitation, cm/yr, m/yr or in/yr
ET = evapotranspiration rate, inch/month
E = efficiency of the irrigation systems.

The leaching factor ranges from 0.05 to 0.30, depending on the crop, the amount of precipitation, and total dissolved solids in the wastewater. For the total dissolved solids of 400 mg/L or more, LR is in the range from 0.1–0.2. The efficiency of the irrigation system is 0.65–0.75 for surface irrigation systems, 0.7–0.8 for sprinklers, and 0.9–0.95 for drip irrigation systems.

6.3.2 Overland-flow systems

Overland-flow is essentially a biological treatment process in which wastewater is treated as it flows over the upper reaches of sloped terraces and allowed to flow across the vegetated surface to runoff collection ditches. Unlike slow-rate systems, overland-flow systems are designed to facilitate the runoff of wastewaters. In order to ensure a runoff, the soil on the slope should be either impervious to water, or slowly permeable, to limit percolation.

A schematic view of the overland-flow treatment is shown in Figure 6.8(a), and a pictorial view of a typical system is shown in Figure 6.8(b). As shown in Figure 6.8(a), there is relatively little percolation involved, either because of an impermeable soil or a subsurface barrier to percolation. Wastewaters are either sprinkler-applied or fan-sprayed or surface-applied (e.g., gated pipe) to the top of the slope. The treatment of the wastewater occurs as the flow runs down the graded land. The slopes normally have a 1–8% gradient and are 30–61 m (100–200 ft) in length, as tabulated in Tables 6.5 and 6.6.

The objectives of overland-flow systems are to achieve high effluent quality by applying the land with pretreated wastewaters and to remove nitrogen, phosphorus, BOD_5, and suspended solids. The pretreatment includes grit and fine screening, primary sedimentation, secondary processes or pond systems. It is also feasible for treating high-strength food processing wastewaters. The collected overland-flow treated water from the ditches is discharged to surface waters.

The primary removal mechanisms for organics and suspended solids are biochemical oxidation, sedimentation, adsorption, and filtration. Nitrogen removal is a combination of plant uptake, denitrification, and volatilization of ammonia nitrogen. The dominant mechanism in a particular site will depend on the forms of nitrogen present in the wastewater, the amount of carbon available, the temperature, and the loading rates and schedules of wastewater application.

Permanent nitrogen removal by the plants is only possible if the crop is harvested and removed from the field. Ammonia volatilization can be significant if the pH of the wastewater is above 7. Nitrogen removals usually range from 75–90%, with the form of runoff nitrogen dependent on temperature and on application loading rates and schedule. Since microbial activities for denitrification occurs at or near the soil surface, the denitrification reactions are adversely affected by cold weather. The same problem also afflicts plant uptake of ammonia, as the majority of crops die off or are in a dormant state in cold winter weather. Less removal of nitrate and ammonium may result from cold temperatures during winter in many non-tropical or non-subtropical regions of the world.

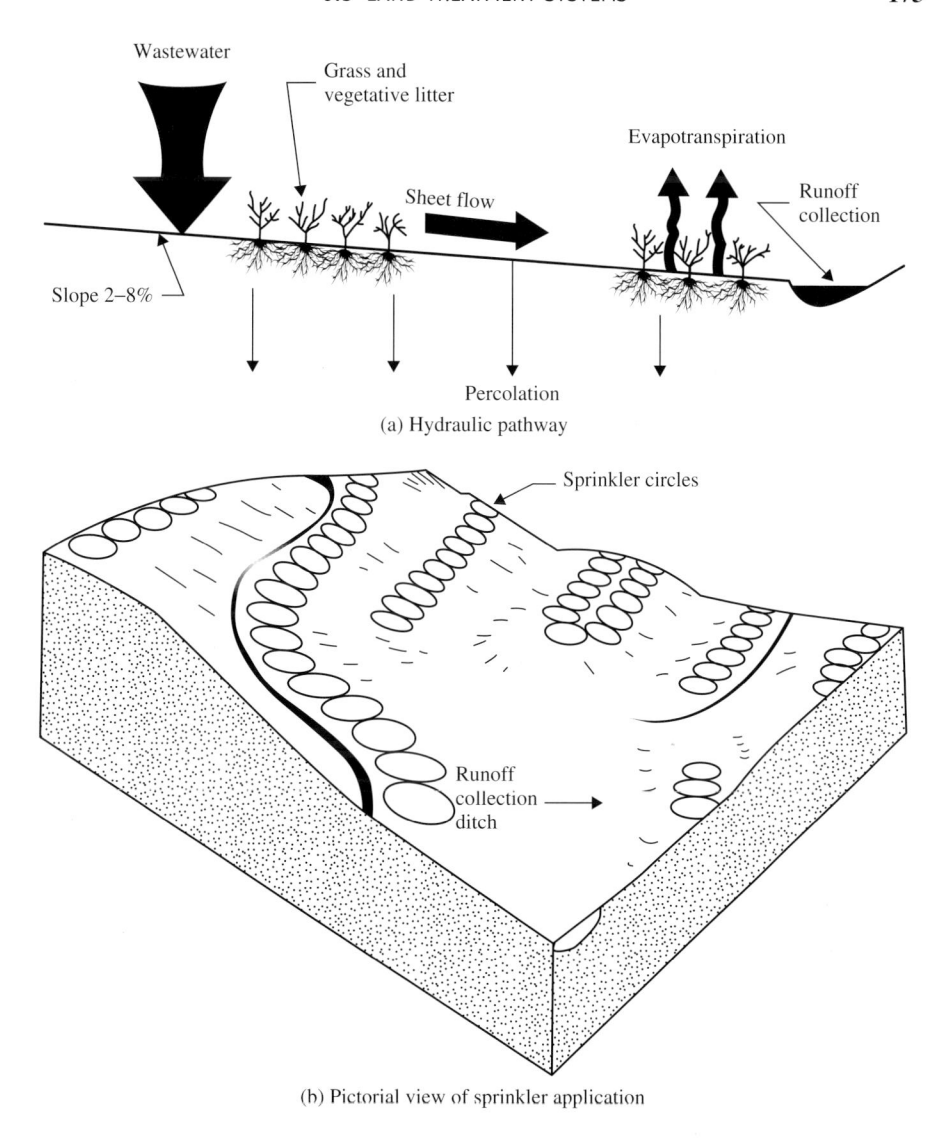

Figure 6.8 (a)(b) A schematic diagram of an overland flow wastewater treatment system. *Source*: U.S. Environmental Protection Agency 1984

Phosphorus is removed by adsorption and precipitation. Treatment efficiencies are somewhat limited because of the limited contact between the wastewater and the adsorption sites within the soil. Phosphorus removals usually range from 50–70% on a mass basis. Increased removals may be obtained by adding coagulants such as alum or ferric chloride to the wastewater just prior to application on the slope.

Overland-flow system design

Soil permeability is an important parameter for designing an overland-flow system, as runoff of wastewater along the slope of the land is required. The best sites for overland-flow systems have soil permeabilities of 0.5 cm/h (0.2 in/h) or less. The high permeability soils can be compacted mechanically to reduce permeability to an acceptable level. Low temperature and rainfall can affect overland-flow systems and, as such, the wastewater application may be curtailed or ceased and the wastewater is stored.

The hydraulic loading rate of wastewater is empirically selected to be from 2–10 cm/d (0.8–4 in/d). Since there is little percolation, the application rate (see the definition below) is more relevant a performance parameter than hydraulic load rate. The two rates are related in the following expression (Reed et al., 1995):

$$L_w = [qp\,(100\ cm/m)]/z \qquad (6.7)$$

where:
- L_w = hydraulic loading rate, cm/d or in/d
- q = application rate per unit width of the slope, m³/h·m or gal/min·ft
- P = application period, h/d (ranging from 6–12 hours per day; 8 hours/day is selected for the purpose of convenience)
- z = distance down-slope, m or ft (ranging from 30–60 m or 100–200 ft; for surface application, the length should be 30–45 m or 100–150 ft)

The application rate is also related to slope length and BOD removal through the following first-order reaction equation (Reed et al., 1995):

$$(C_z - C)/C_0 = A\ \exp[(-kz)/q^n] \qquad (6.8)$$

where:
- C_z = BOD$_5$ concentration of surface flow at a distance (z) down-slope, mg/L
- C = background BOD$_5$ at the end of the slope
- C_0 = BOD$_5$ concentration of applied wastewater, mg/L
- z = distance down-slope, m or ft
- A = empirically determined coefficient dependent on the value of q
- q = application rate, m³/h·m or gal/min·ft
- n = empirically determined exponents (<1)
- k = empirically determined rate constant

For overland-flow treatment of high-strength industrial wastewater, such as some food processing wastewaters, the BOD$_5$ loading must be considered. In order to avoid excessive amounts of organic loading on the slope, which leads to anaerobic activities due to exhausting the oxygen at or near the soil surface, the BOD$_5$ loading rate is controlled at:

$$L_{BOD} = B\,L_w C_0 \qquad (6.9)$$

where:

L_{BOD} = BOD$_5$ loading rate, kg/ha·d or lb/ac·d
B = conversion factor = 0.1 for metric units or 0.225 for imperial units
L_w = hydraulic loading rate, cm/d or in/d

= qPWm/z
q = application rate, m³/h·m or gal/min·ft
P = application period, h/d
W = width of application slope, m or ft
z = distance down-slope, m or ft
m = conversion factor = 100 cm/m or 96.3 for imperial units
C_0 = BOD$_5$ concentration of applied wastewater, mg/L.

When BOD$_5$ concentration of wastewater exceeds 800 mg/L, the wastewater has been diluted in order to avoid excessive anaerobic activities.

Land requirements for overland-flow systems depend on the flow, the application rate, and the application period. The required surface area for overland-flow treatment is:

$$A_s = Q\,z/(q\,P\,C) \qquad (6.10)$$

where:

A_s = field surface area required, ha or ac
Q = wastewater flow rate, m³/d or gal/min
z = distance down-slope or slope length, m or ft
q = application rate, m³/h·m or gal/min·ft
P = application period, h/d
C = conversion factor = 10,000 m²/ha or 726 for imperial units.

If wastewater storage is anticipated, the field area becomes:

$$A_s = (365Q + V_s)/(D\,L_w C') \qquad (6.11)$$

where:

A_s = field surface area required, ha or ac
V_s = net loss or gain in storage volume due to precipitation, evaporation, and seepage, m³/yr or ft³/yr

D = number of operating days per year
L_w = design hydraulic loading, cm/d or in/d
C' = conversion factor = 100 for metric units or 3,630 for imperial units.

If BOD$_5$ loading is a limiting factor, the field area is expressed as:

$$A_s = (C_0 C'' Q_a)/L_{LBOD} \qquad (6.12)$$

where:

A_s = field surface area required, ha or ac
C_0 = BOD$_5$ concentration of applied wastewater, mg/L
C'' = conversion factor = 0.1 for metric units and 6.24×10^{-5} for imperial units
L_{LBOD} = limiting BOD$_5$ loading rate = 100 kg/ha·d or 89 lb/ac·d
Q_a = design flow rate to the overland-flow site, m^3/d or ft^3/d.

6.3.3 Rapid-infiltration systems

The objective of the application of rapid-infiltration systems is to recharge or store renovated water in the underground aquifer and, in some cases, to recharge surface waters using under-drains or wells to channel the water to the adjacent surface water body. In rapid-infiltration land treatment, most of the applied wastewater percolates through the soil, and the treated effluent drains naturally to surface waters or joins the groundwater. The wastewater is applied to moderately and highly permeable soils (such as sands and loamy sands) by spreading in basins or by sprinkling, and it is treated as it travels through the soil matrix. Vegetation is not usually planned, but there are some exceptions, and emergence of weeds and grasses usually does not cause problems.

The schematic view in Figure 6.9(a) shows a typical hydraulic pathway for rapid infiltration systems. A much greater portion of the applied wastewater percolates to the groundwater than with SR land treatment. There is little or no consumptive use by plants. Evaporation ranges from about 0.6 in/yr (2 ft/yr) for cool regions, to 2 in/yr (6 ft/yr) for hot arid regions. This is usually a small percentage of the hydraulic loading rates.

In many cases, recovery of renovated water is an integral part of the system. This can be accomplished using under-drains or wells, as shown in Figure 6.9(b). In some cases, the water drains naturally to an adjacent surface water body (Figure 6.9(c)). Such systems can provide a higher level of treatment than most mechanical-intensive systems for discharging to the same surface water.

6.3 LAND TREATMENT SYSTEMS

(a) Hydraulic pathway

(b) Recovery pathway

(c) Natural drainage into surface waters

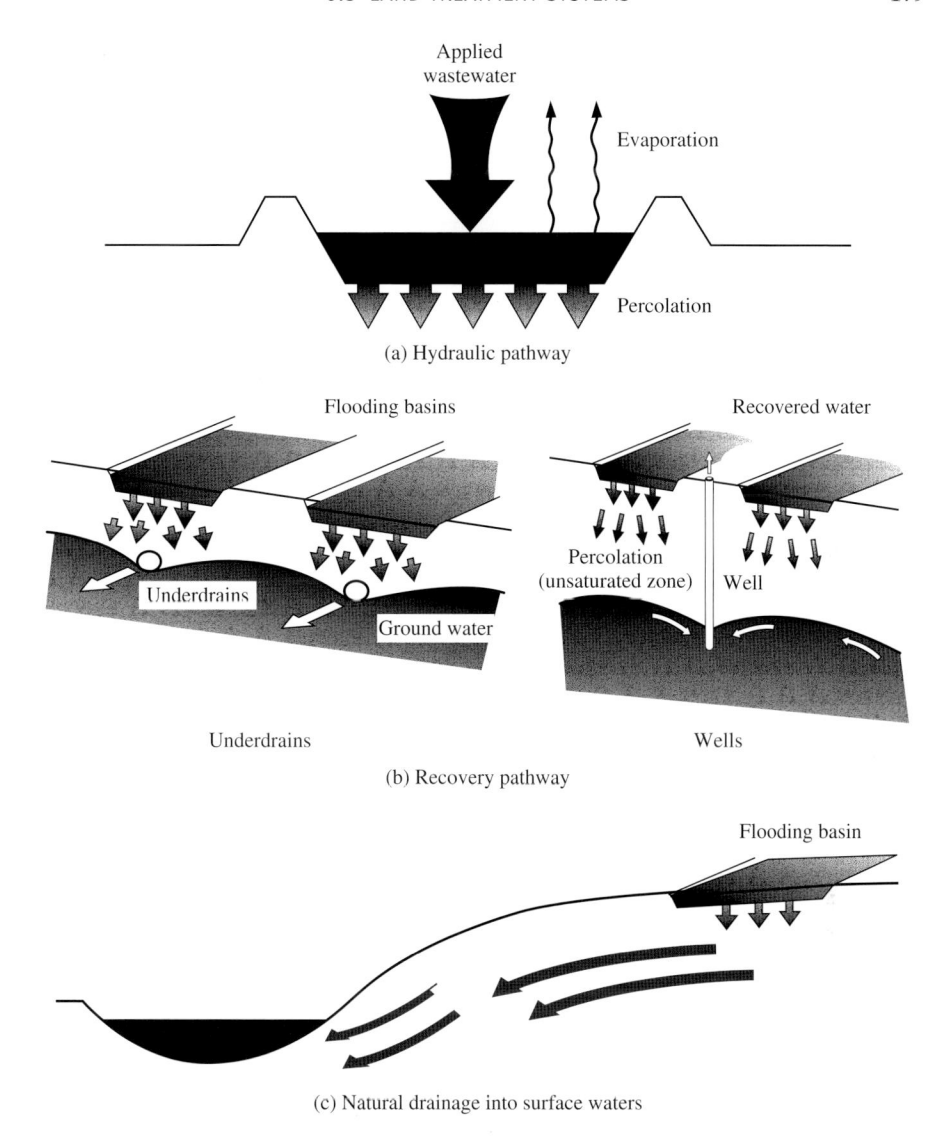

Figure 6.9 (a)(b)(c) A schematic diagram of a rapid infiltration wastewater treatment system. *Source*: U.S. Environmental Protection Agency 1984

Removals of wastewater constituents by the filtering and straining action of the soil are superb. Suspended solids, BOD_5, and fecal coliforms are almost completely removed. Nitrification of the applied wastewater is essentially complete when appropriate hydraulic loading cycles are used. Thus, for communities that have ammonia standards in their discharge requirements, rapid-infiltration can provide an effective way to meet these standards. Generally, nitrogen removal averages 50% unless specific operating procedures are established to maximize denitrification. These procedures include optimizing the application cycle, recycling the portions of the renovated water that contain high nitrate concentrations, reducing the infiltration rate, and supplying an additional carbon source. Using these procedures in soil column studies, average nitrogen removals of 80% have been achieved.

Phosphorus removals can range from 70% to 99%, depending on the physical and chemical characteristics of the soil. As with slow-rate systems, the primary removal mechanism is adsorption, with some chemical precipitation, so the long-term capacity is limited by the mass and the characteristics of the soil in contact with the wastewater. Removals are related also to the residence time of the wastewater in the soil, the travel distance, and other climatic and operating conditions.

Rapid-infiltration system design

The design of annual hydraulic loading rate of a rapid-infiltration system is based on the permeability of the soil or the effective hydraulic conductivity of the soil media the wastewater infiltrates. The rate is expressed as follows (Metcalf & Eddy, 1991):

$$L_w = (IR)(OR) \, F \, C''' \tag{6.13}$$

where:
 L_w = hydraulic loading rate, ft/year or cm/yr
 IR = infiltration rate, in/h, cm/h
 OD = number of operating days per year, day/yr

F = application factor (10–15% of the minimum measured infiltration rate if basin infiltration is used; 4–10% of the conductivity of the most restrictive soil layer if vertical hydraulic conductivity measurements)

 C''' = conversion factor = 2 for imperial units and 24 for metric units.

Application of a rapid-infiltration system to wastewater treatment is not continuous but in cycles. The purpose of non-application periods (dry periods) is to allow the land to be re-aerated and to decompose the accumulated organic matter in the soil. The application rate thus can be calculated based on the hydraulic rate, operating cycle time, and application period. The land requirement of a rapid-infiltration system is obtained by dividing the annual wastewater flow rate by the design annual hydraulic loading rate in Equation (6.13).

6.4 Wetland systems

Natural wetlands (e.g., swamps, bogs, marshes, fens, sloughs, etc.) have long been recognized as providing many benefits, including: food and habitat for wildlife; water quality improvement; flood protection; shoreline erosion control; and opportunities for recreation and aesthetic appreciation. Many of these same benefits have been realized by projects across the country which involve the use of wetlands in wastewater treatment.

Wetlands are constructed as either surface flow (see Figure 6.10) or sub-surface flow systems (see Figure 6.11). Both types of wetlands treatment systems are typically constructed in basins or channels with a natural or constructed sub-surface barrier to limit seepage (US EPA, 1993). Surface flow systems require more land, but they are generally easier to design, construct and maintain. They consist of shallow basins with emergent and submergent

Figure 6.10 A surface flow wetland system in the USA. *Source*: USEPA (see plate section for color version)

Figure 6.11 A subsurface flow wetland system in the US. *Source*: USEPA (see plate section for color version)

wetland plants that tolerate saturated soil and aerobic conditions. Water flows in one end of the basin, moves slowly through, and is released at the other end. These systems provide habitat and public access.

Subsurface flow systems are designed to create subsurface flow through a permeable medium, keeping the water being treated below the surface and, thereby, helping to avoid the development of odors and other nuisance problems. Such systems have also been referred to as "root-zone systems", "rock-reed-filters" and "vegetated submerged bed systems." The choice of media used (typically soil, sand, gravel or crushed rock) greatly affects the hydraulics of the system. These systems demonstrate higher rates of contaminant removal than surface flow wetlands.

Both systems utilize the roots of plants to provide substrate for the growth of attached bacteria that use the nutrients present in the effluents, and for the transfer of oxygen. Bacteria do the bulk of the work in these systems, although

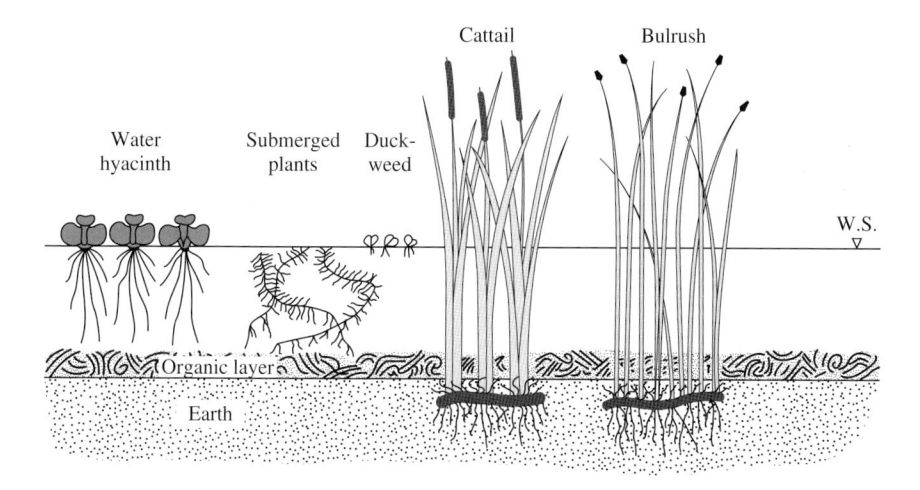

Figure 6.12 A common aquatic species for constructed wetlands. *Source*: U.S. Environmental Protection Agency 1993

there is some nitrogen uptake by the plants. The surface flow system most closely approximates a natural wetland. Typically, these systems are long, narrow basins with depths of less than two feet, which are planted with aquatic vegetation such as bulrush *(Scirpus* spp.) or cattails *(Typha* spp.) (see Figure 6.12). The shallow subsurface flow systems use a gravel or sand medium, approximately 18 inches deep, which provides a rooting medium for the aquatic plants and through which the wastewater flows.

Organic and inorganic matters are removed from the wastewater through several mechanisms. Processes of sedimentation, microbial degradation, precipitation, and plant uptake remove most contaminants. Organic compounds can be broken down for consumption by microorganisms in a wetland system. This biodegradation removes the organic compounds from water, as they provide energy for the organisms. Organics can also be degraded when taken up by plants, and they can also sorb to surfaces in the wetland, usually to plant debris. Organic compounds containing nitrogen sorb to the surface, and organic nitrogen is converted to ammonia. Ammonia can volatilize, be exchanged with other cations in the sediment, or be nitrified if oxygen is present. Nitrate is the form of nitrogen taken up by plants, so emergent plants use it during the growing season. Excess nitrate in an anaerobic system is reduced to N_2 and N_2O gases as a result of denitrification, the main mechanism of nitrate removal.

The design considerations for constructed wetlands systems are varied and site-dependent. Food and agricultural wastewater treatment systems are most concerned with the reduction of suspended solids, organic matter, pathogens, phosphates, ammonia, and organic nitrogen. Some system designs anticipate exactly what kinds of contaminants the wetlands will receive, and at what levels, while others face variable and unpredictable wastewater streams.

Food and agricultural wastewater destined for wetlands treatment often travels through a treatment train, although, in some cases, wastewater is released directly into a wetland system. The initial step is usually passage through a traditional wastewater treatment plant, where excess ammonia is removed, followed by a sedimentation chamber, where any remaining suspended material is removed. Depending on the levels of fecal coliforms and the requirements for effluent contaminant levels, the water may be disinfected with chlorine before release into the constructed wetland system. If the water is to be discharged into a waterway, the minimum contaminant criteria may be different than for a system in which the wetlands are the final destination for the water.

Engineered wetlands for other kinds of wastewater may also consist of a series of treatment steps that have been built according to the expected flow and loading rates. In general, the heavier the load a system receives, the larger the wetlands system will need to be in order to remove contaminants effectively. The heavier load could be a large volume of water discharged into the system, or volumes with higher concentrations of contaminants.

A series of lined settling and aeration ponds, or lagoons, may be the initial step in treatment, followed by release into the actual wetland. Wetland designs can vary from more traditional systems, with populations of native plants, to aerobic systems that function without aquatic plants and treat waste primarily with added bacteria. An aerobic system may use aquatic plants in a final polishing step.

The principal design parameters for constructed wetland for wastewater treatment include hydraulic detention time, basin depth, basin geometry, BOD_5 loading rate, and hydraulic loading rate (Metcalf & Eddy, 1991). Table 6.7 provides the typical ranges of design parameters for constructed wetland.

Table 6.7 Typical design parameters for constructed wetlands

Design factor	Surface water flow	Subsurface water flow
Minimum surface area	23–115 ac/mgd	2.3–46 ac/mgd
Maximum water depth	Relatively shallow	Water level below ground surface
Bed depth	Not applicable	12.30 m
Minimum hydraulic residence time	7 days	7 days
Maximum hydraulic loading rate	0.81–4.07 cm/d	2.03–40.5 cm/d
Minimum pretreatment	Primary (secondary optional)	Primary
Range of organic loading as BOD	4.1–8.2 kg/ac·d	0.82–63.6 kg/ac·d

Source: U.S. Environmental Protection Agency 1992.

Simplified mathematical expressions for wetland systems are available if we view the systems as aerobic attached-growth biological reactors with plug-flow kinetics for BOD$_5$ and nitrogen removal. The basic equation for both systems of wetland treatment (Reed et al., 1995) is:

$$C_e/C_0 = \exp(-k_T t) \quad (6.14)$$

where:
- C_e = effluent pollutant concentration, mg/L
- C_0 = influent pollutant concentration, mg/L
- k_T = temperature-dependent first-order reaction rate constant, d^{-1}
- t = hydraulic residence time, d.

The hydraulic residence time in the wetland can be computed as

$$T = (L\ W\ yn)/Q \quad (6.15)$$

where:
- L = length of the wetland cell, m or ft
- W = width of the wetland cell, m or ft
- y = depth of water in the wetland cell, m or ft
- n = porosity, formed by vegetation, little, media, roots, and other solids depending on the system, expressed as a percentage
- Q = the average flow through the wetland (Q = (Q_{in} + Q_{out})/2), m^3/d or ft^3/d.

The surface area of the wetland is given by A_s = LW.

One common problem associated with surface flow wetland systems is mosquitoes when certain pockets of the wetland hold still waters. This problem can be controlled by biological means. There are certain fish species (e.g., mosquito fish, *Gambusia affinis*) that feed upon mosquito's egg, larva, pupa; they can be introduced into the wetland system provided that the DO level in the wetland is above 1 mg/L in order to maintain fish populations.

6.5 Floating aquatic plant systems

Aquatic plant systems are engineered and constructed systems that use aquatic plants in the treatment of various sources of wastewater. They are designed to achieve a specific wastewater treatment goal. Aquatic plant systems can be divided into two categories:

- Systems with floating aquatic plants, such as water hyacinth, duckweed, pennywort;
- Systems with submerged aquatic plants such as waterweed, water milfoil, and watercress.

The use of aquaculture as a means of treating wastewater involves both natural and artificial wetlands and the production of algae and higher plants (submersed and immersed) to remove contaminants such as nitrogen compounds, BOD_5, hydrocarbons, and heavy metals from the wastewater. Floating aquatic plants such as water hyacinth *(Eichhornia crassipes)* and duckweed *(Lemna spp.)* appear to be among the most promising aquatic plants for the treatment of wastewater, and they have received the most attention in this regard. Other plants are also being studied, however, among them seaweeds and alligator weed.

These systems are basically shallow ponds covered with floating plants that detain wastewater for at least one week. The main purpose of the plants in these systems is to provide a suitable habitation for the bacteria that remove the vast majority of dissolved nutrients. The design features of such systems are summarized in Table 6.8.

These technologies are useful in areas where suitable plants are readily available. In areas where they are not, any introduction of plants species must be undertaken with caution in order to minimize the possibility of creating nuisance growth conditions. Even introducing them into constructed enclosures should be done carefully, and with the foreknowledge that there is a strong likelihood that they will enter natural water systems (especially as they must be harvested from the treatment systems and disposed of).

Table 6.8 Performance of four different wastewater effluent treatment systems using water hyacinth

Source	BOD reduction	COD reduction	TSS reduction	Nitrogen reduction	Phosphorus reduction
Secondary effluent	35%	n/a	n/a	44%	74%
Secondary effluent	83%	61%	83%	72%	31%
Raw wastewater	97%	n/a	75%	92%	60%
Secondary effluent	60-79%	n/a	71%	47%	11%

[a]*Source*: U.S. Environmental Protection Agency 1976.

6.5 FLOATING AQUATIC PLANT SYSTEMS

Advantages include:

- The cost of plant seeding and wetlands is very low and, in most cases, negligible.
- These technologies are traditional, rudimentary, and easy to implement – ideal for rural areas.
- Wetland systems are easy to build, simple to operate, and require little or no maintenance.
- Most small-scale wetland treatment systems require relatively small land areas.
- Wetland technologies reduce nutrient contamination of natural systems.
- Heavy metals absorbed by the plants in wetland treatment systems are not returned to the water.
- Water hyacinth-based and other wetland systems produce plant biomass that can be used as a fertilizer, animal feed supplement, or as a source of methane for fuel.

Disadvantages include:

- In some places, plant seeds may not be readily available.
- Temperature (climate) is a major limitation, since effective treatment is linked to the active growth phase of the immersed (surface and above) vegetation.
- Herbicides and other materials toxic to the plants can affect their health and lead to a reduced level of treatment.
- Duckweed is prized as food by waterfowl and fish, and it can be seriously depleted by these species.
- Winds may blow duckweed to the windward shore unless windscreens or deep trenches are employed.
- Plants die rapidly when the water temperature approaches freezing point; therefore, greenhouse structures may be necessary in cooler climates.
- Water hyacinth is sensitive to high salinity, which restricts the removal of potassium and phosphorus to the active growth period of the plants.
- Metals such as arsenic, chromium, copper, mercury, lead, nickel and zinc can accumulate in water hyacinth plants and limit their suitability as fertilizer or feed materials.

- Water hyacinth plants may create small pools of stagnant surface water that can serve as mosquito breeding habitat; this problem can generally be avoided by maintaining mosquito fish (*Gambusia affinis*) or similar fishes in the system.
- The spread of water hyacinth must be closely controlled by barriers, since the plant can spread rapidly and clog previously unaffected waterways.
- Water hyacinth treatment may prove impractical for large-scale treatment plants because of the land area required.
- Evapotranspiration in wetland treatment systems can be 2–7 times greater than evaporation alone.
- Harvesting the water hyacinth or duckweed plants is essential to maintain high levels of system performance.

The water hyacinth is a perennial, free-floating freshwater aquatic macrophyte with rounded, upright, thick, waxy, gloss-green leaves and spikes of lavender flowers (see Figure 6.13). It is native to South America and found naturally in waterways, bayous, and other backwaters in temperate and tropical regions. The water hyacinth is considered as one of the worst weeds in the world – aquatic or terrestrial – because of its fast growth that tends to clog waterways for boat traffic and prevent sunlight and oxygen from getting into the water. It thrives in nitrogen-rich environments, and consequently it does extremely well in raw and partially treated wastewaters.

When water hyacinth is used for effluent treatment, wastewater is passed through a water-hyacinth-covered basin, where the plants remove nutrients, suspended solids, heavy metals, and other contaminants. Batch treatment and flow-through systems, using single and multiple lagoons, are used. Because of its rapid growth rate and inherent resistance to insect predation and disease, water hyacinth plants must be harvested from these systems. While many uses of the plant material have been investigated, it is generally recommended as a source of methane when anaerobically digested. Its use as a fertilizer or soil conditioner (after composting), or as an animal feed, is often not recommended, owing to its propensity to accumulate heavy metals. The plant also has a low organic content (it is primarily water) and, when composted, it leaves behind little material with which to enrich the soil.

Duckweeds (*Lemna sp., Spirodela sp., and Wolffia sp.*) are the smallest flowering plants. They are small, green freshwater plants with fronds from one to a few millimeters and a short root, usually less than one centimeter in length, and they grow as small colonies of plants floating on the surfaces of quiet bodies of water. Their growth can be extremely rapid, given the proper conditions. These plants are almost all leaf, having essentially no stem tissue and only one, or a few, very fine roots. Figure 6.14 shows three species of duckweeds.

Figure 6.13 Water hyacinth in Florida waterway. *Source*: Willey Durden, USDA ARS (retired) (see plate section for color version)

In nature, duckweeds serve as food for many species of fish and aquatic birds. In the United States, crayfish are often released in irrigated rice fields in rice-growing areas to control weeds (often duckweeds). Since duckweeds are high in proteins and also β-carotene, harvested duckweeds can be used as feeds for grass carp, sheep, chicken, and tilapia. Duckweeds can tolerate and grow under a wide range of conditions, including on water polluted with high concentrations of bacteria and some agricultural wastes. These characteristics have brought the duckweeds to the attention of environmental engineers and aquaculturists alike.

Water hyacinth systems

They are predominant floating aquatic plant systems for wastewater treatment. Three types of the systems exist today for BOD_5 and nutrient removals: aerobic non-aerated, aerobic aerated, and facultative anaerobic (Metcalf & Eddy, 2001).

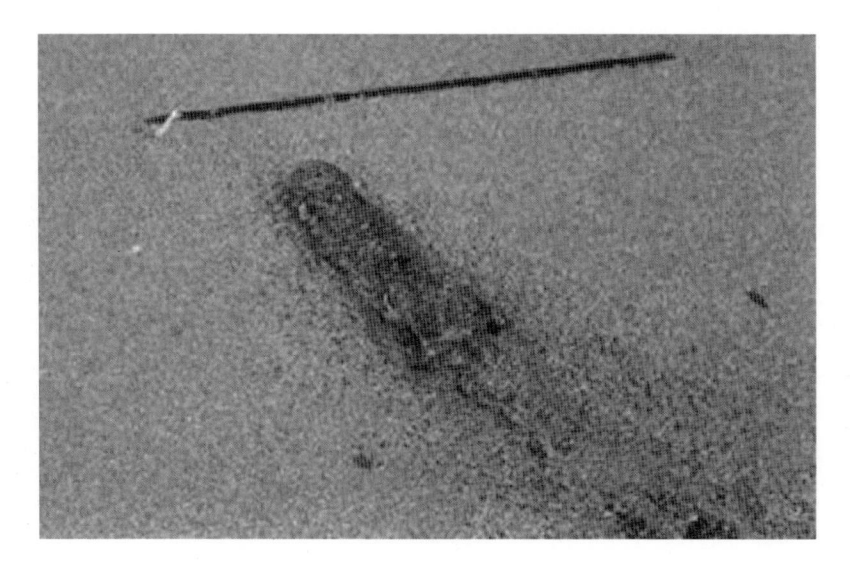

Figure 6.14 Three species of native duckweeds in Florida waterway. *Source*: Willey Durden, USDA ARS (retired) (see plate section for color version)

These types of water hyacinth systems mirror the types of stabilization ponds used for wastewater treatment. Not surprisingly, floating aquatic plants for wastewater treatment, such as water hyacinth, thrive in natural or artificial ponds and constructed wetlands. However, in terms of geographic location, the use of water hyacinth has been limited to warm weather regions, because of the sensitivity of the plant to freezing conditions (see Figure 6.15). Water hyacinth systems have most often been used either for removing algae from oxidation pond effluents (by blocking the sunlight to reach the water) or for nutrient removal following secondary treatment.

Design criteria for wastewater treatment using water hyacinth include: the depth of the lagoons, which should be sufficient to maximize root growth and the absorption of nutrients and heavy metals; detention time; the flow rate and volume of effluent to be treated; and the desired water quality and potential uses of the treated water. Land requirements for pond construction are approximately $1 \text{ m}^2/\text{m}^3/\text{day}$ of water to be treated. Phosphorus reductions obtained in such systems range between 10% and 75%, and nitrogen reductions between 40% and 75% of the influent concentration.

Duckweed systems

Duckweed systems have been used successfully to improve the effluent quality from facultative stabilization ponds by reducing algae population, BOD_5, and

6.5 FLOATING AQUATIC PLANT SYSTEMS

Figure 6.15 A map of water hyacinth growing region in the USA. *Source*: U.S. Environmental Protection Agency 1993

nutrients. Duckweeds are sensitive to wind and may be blown in drifts to the leeward side of the pond, resulting in exposure of a large surface area that may be prone to algal blooming. The regions in the USA that are suitable for duckweed systems are shown in Figure 6.16. Redistribution of the plants to cover the surface requires manual labor. Also, piles of decomposing plants can result in the production of odors. Drift control, such as floating baffles, can be used

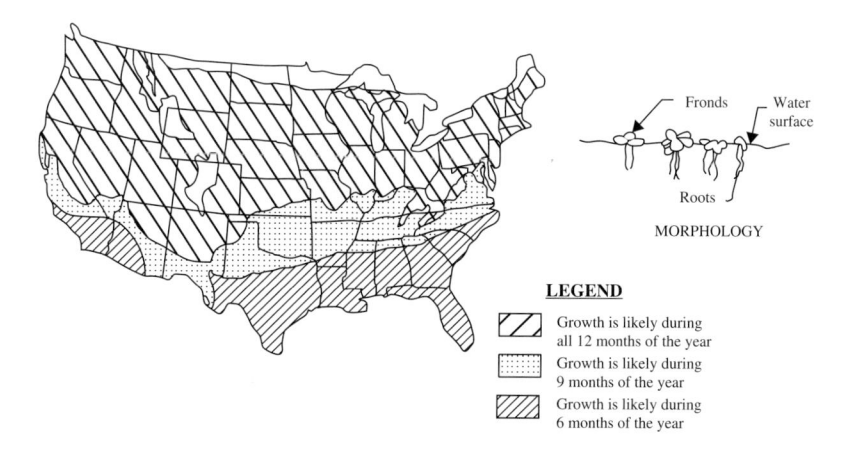

Figure 6.16 A map of duckweed growing region in the USA. *Source*: U.S. Environmental Protection Agency 1993

Table 6.9 Performance of duckweed system for treating facultative pond effluents in several US locations

Location	BOD$_5$, mg/L influent	BOD$_5$, mg/L effluent	TSS, mg/L influent	TSS, mg/L effluent	Depth, m	Detention time, days
Biloxi, MS	30	15	155	12	2.4	21
Collins, MS	33	13	36	13	0.4	7
Sleep eye, MN	420	18	364	34	1.5	70

Source: U.S. Environmental Protection Agency 1988.

to divide the surface area into smaller cells, thus reducing the extent of open surface area exposed to blowing wind.

Similar to water hyacinth systems, design criteria of duckweed systems include hydraulic detention time, water depth, pond geometry, BOD$_5$ loading, and hydraulic loading rate. Typical treatment results from several US locations involving duckweed wastewater treatment systems for effluents from facultative stabilization ponds are provided in Table 6.9.

6.6 Further reading

Crites, R.W., Reed, S.C., & Bastian, R. (2000). *Land Treatment Systems for Municipal and Industrial Wastes*. McGraw-Hill Professional, New York, NY.

Mara, D. (2004). *Domestic Wastewater Treatment in Developing Countries*. Earthscan, London, UK.

Metcalf & Eddy, Inc. (Tchobanoglous, G. & Burton, F.L.). (1991). *Wastewater Engineering, Treatment, Disposal, and Reuse*. 3rd edition. McGraw-Hill, New York, NY.

Reed, S. C., Crites, R. W., & Middlebrooks, E. J. (1995a). *Natural Systems for Waste Management and Treatment*. 2nd edition. McGraw-Hill, Inc., New York, NY.

6.7 References

Gomes de Sousa, J.M. (1987). Wastewater stabilization lagoon design criteria for Portugal. *Water Science and Technology* **19**, 7–16.

Jayawardane, N.S., Blackwell, J., Cook, F.J., Nicoll, G., & Wallet, D. (1997). *Final report on pollutant removal by the FILTER system during the period November 1994 to November 1996*, Prepared for Griffith City Council. Report No. 97. CSIRO Division of Land and Water Consultancy, Canada.

6.7 REFERENCES

Mara, D.D., Alabaster, G.P., Pearson, H.W. & Mills S.W. (1992). *Waste Stabilisation Ponds – A Design Manual for Eastern Africa.* Lagoon Technology International Ltd, Leeds, UK.

Mara, D.D. & Pearson, H.W. (1998). *Design Manual for Waste Stabilization Ponds in Mediterranean Countries.* Lagoon Technology International, Leeds, UK.

Metcalf & Eddy, Inc. (Tchobanoglous, G. & Burton, F.L). (1991). *Wastewater Engineering, Treatment, Disposal, and Reuse.* 3rd edition. McGraw-Hill, New York, NY.

Pano, A. & Middlebrooks, E.J. (1982). Ammonia nitrogen removal in facultative wastewater stabilization ponds. *Journal of the Water Pollution Control Federation* **54**, 344–351.

Paranychianakis, N.V., Angelakis, A. N., Leverenz, H. & Tchobanoglous, G. (2006). Treatment of wastewater with slow rate systems: a review of treatment processes and plant functions. *Critical Review in Environmental Science and Technology* **36**, 187–259.

Park, J.B.K., Craggs, R.J. & Shilton, A.N. (2011). Wastewater treatment high rate algal ponds for biofuel production. *Bioresource Technology* **102**, 35–42.

Pavia, E.H. & Tyagi, A.D. (1972). *Safe Economical Rescue of Transport Water in the Fish Meal and Oil Industry.* ASME, New York, NY.

Pescod M.B. & Mara D.D. (1988). Design, operation and maintenance of wastewater stabilization ponds. In: Pescod, M.B. & Arar, A. (eds). *Treatment and Use of Sewage Effluent for Irrigation.* Butterworths, Sevenoaks, Kent, UK.

Racault, Y., Boutin, C. & Sequin, A. (1995). Waste stabilization ponds in France: a report on fifteen years of experience. *Water Science and Technology* **31**, 91–101.

Reed, S.C. & Bastian, R.K. (1985). Wetlands for wastewater treatment: an engineering perspective. In: Godfrey, P.J. (ed.), *Ecological Considerations in Wetlands Treatment of Municipal Wastewaters.* Van Nostrand Reinhold, New York, NY.

Reed, S.C., Crites, R.W. & Middlebrooks, E.J. (1995b). *Natural Systems for Waste Management and Treatment.* 2nd edition. McGraw-Hill, Inc., New York, NY.

Townshend, A.R. & Knoll, H. (1987). *Cold Climate Sewage Lagoons.* Report EPS 3/NR/1. Environmental Canada, Ottawa, Canada.

US Environmental Protection Agency (1976). *Innovative and Alternative Technology Assessment Manual.* EPA 430/9-78-009. US EPA, Washington D.C.

US Environmental Protection Agency (1981). *Process Design Manual for Land Treatment of municipal Wastewater, Supplements on Rapid Infiltration and Overland Flow.* EPA 625/1-81-013a. US EPA, Center for Environmental Research Information, Cincinnati, OH.

US Environmental Protection Agency (1983). *Operational Manual for Stabilization Ponds.* US EPA, Washington D.C.

US Environmental Protection Agency (1984). *Process Design Manual for Land Treatment of municipal Wastewater.* EPA 625/1-81-013. US EPA, Center for Environmental Research Information, Cincinnati, OH.

US Environmental Protection Agency (1988). *Design Manual: Constructed Wetlands and Aquatic Plant Systems for Municipal Wastewater Treatment.* EPA

625/1-88-022. US EPA, Center for Environmental Research Information, Cincinnati, OH.

US Environmental Protection Agency (1992). *Wastewater Treatment/Disposal for Small Communities*. EPA-625/R-92-005. US EPA, Center for Environmental Research Information, Cincinnati, OH.

US Environmental Protection Agency (1993). *Constructed Wetlands for Wastewater Treatment and Wildlife Habitat*. EPA 832/R-93-005. US EPA, Center for Environmental Research Information, Cincinnati, OH.

WHO (1987). *Wastewater Stabilization Ponds, Principles of Planning & Practice*. WHO Technical Publication 10. Regional Office for the Eastern Mediterranean, Alexandria, Egypt.

Zachritz, W.H., Mexal, J. & Sammis, T. (2001). Land application of wastewater in arid regions: the challenge of balancing plant water requirements and nitrogen uptake. http://hydrology1.nmsu.edu/hydrology/wastewater/Waste-water-paper.htm

7
Sludge treatment and management

In essence, wastewater treatment involves a major operation of separation, which concentrates and converts suspended and soluble nutrients into a settleable form that can be separated from the bulk of the liquid. The settleable byproducts of wastewater treatment from preliminary treatment, primary treatment, and secondary treatment processes and/or advanced treatment processes are called sludges. The higher the degree of treatment of food and agricultural wastewater, the more sludge is created in the processes involved, and this sludge has to be appropriately handled unless it is going to be used for land applications or polishing lagoons.

The complexity of sludge, due mainly to the characteristics of raw wastewater and treatment processes involved, often means that sludge management is a costly operation in wastewater treatment. The proper utilization and disposal of sludge is one of the most critical issues facing wastewater treatment plants today. Nearly all wastewater treatment plants face the problem of storing and disposing sludges. Landfill costs are skyrocketing, incineration permits are expensive and difficult to obtain, and land application is limited by the availability of permitted land.

Additionally, the amount of sludge produced in wastewater treatment is large, even though 97% of it is primarily water trapped in the solids. Addition of chemicals in physicochemical treatment processes of wastewater treatment increases the amount of sludge produced in the treatment plant. Given the difficulty associated with direct disposal of this enormous watery bulk of waste materials to either landfill or water ways, it is not surprising that the main task of sludge treatment is to reduce the quantity of sludge through removal of water.

Sludge also needs to be "stabilized" by converting organic solids into harmless inert forms so that the treated sludge can be handled or used as soil conditioners without causing a nuisance or health hazards. Several basic operations of sludge

---

Food and Agricultural Wastewater Utilization and Treatment, Second Edition. Sean X. Liu.
© 2014 John Wiley & Sons, Ltd. Published 2014 by John Wiley & Sons, Ltd.

treatment are commonly used in food and agricultural wastewater treatment: thickening, stabilization, conditioning, dewatering, reduction, and disposal.

7.1 Sludge quality and characteristics

Sludge is rather complex material, both in the sense of its composition and the characteristics of its fluid dynamics, which are shaped by its moisture content and interactions between water molecules and other solid particles. In general, when the moisture content of the sludge exceeds 90% by mass or weight, the sludge behaves as a Newtonian fluid, whereas below 90%, it behaves as a non-Newtonian fluid, in some ways similar to how polymers behave (both biopolymers such as foods and petroleum-based polymers).

The water in sludge is held either as free water or as bound water. It is estimated that in sludge with 95% moisture content, approximately 70% of it is free water, while the remaining is bound. The free water molecules are easier to separate from the sludge. Among the 30% that is bound water, 20% is present as a part of flocs or aggregates, 8% is chemically bound, and 2% is capillary water. As a rule, the intimately bound water, such as capillary water and chemically bound water, requires more energy to remove than does free water. Free water in sludge can be removed by gravity thickening. However, this type of moisture reduction may not be enough for sludge handling and further processing.

Chemicals and mechanical means are used to "condition" the sludge and make it more amenable to dewatering. Mechanical devices break up the structure of the sludge to release trapped water from flocs, while chemicals such as coagulants such as $FeCl_3$, alum or polyelectrolytes are used to break up and alter the structure and composition of the sludge, resulting in improved dewaterability.

The application of basic sludge treatment techniques, whether stabilization or thickening, depends on the sludge quality and characteristics. Different industries produce different types of sludges that affect sludge treatment options and, ultimately, disposal. Sludges with gross organic solids, such as offal from meat processing operations, or pathogens, heavy metals and other toxic materials from pharmaceutical, chemical, and metal processing industries, may have few disposal options. This issue needs to be carefully considered by sludge management and treatment personnel and designers during the planning phase of their work.

The characteristics of solids and sludges from wastewater treatment processes also vary with the processes or unit operations of wastewater treatment plants. The retained solids from screens and grit are large-sized inorganic or organic materials, such as vegetable leaves or other food debris that are large enough to be removed on bar racks. Depending on the nature of the food processing and the season of the year, scum from the floatation process of primary and secondary settling tanks could consist of grease, vegetable, fruit skim, animal fat, and floatable food wastes.

Primary sludge from the bottom of the settling tank appears gray in color and is slimy, and it emits offensive odors, while sludge from chemical precipitation with metal salts may be dark in color. Sludge from secondary settling varies with the nature of the biological process employed; sludge from an activated sludge process has a brownish, flocculant appearance and it smells far less offensive than does primary sludge if the activated sludge process is well oxygenated. Sludge sloshed down from a trickling filter is brownish, flocculant and relatively inoffensive. Digested sludges, regardless of whether they are aerobic or anaerobic, are dark brown to black and do not have an offensive odor if treated thoroughly.

Since most wastewater sludges are comprised of water, their properties depend on their water content. For example, once the percentage of solids and the specific gravity of a sludge are known, the volume of the sludge can be estimated. The mass balances used to describe the basic unit operations of sludge treatment can be established based on the solids in the sludges.

The characteristics of stabilized or raw sludge that affect its suitability for land application and beneficial use include organic content, nutrients, pathogens, metals, and toxic organics. The fertilizer value of sludge, if evaluated and found to be suitable, is based mainly on its contents of nitrogen, phosphorus, and potassium. In some food and agricultural wastewater treatment operations, sludge may not have sufficient phosphorus and potassium contents to provide for good plant growth, and trace amounts of inorganic compounds in the sludge may spur or stunt the growth of plants. Heavy metals in sludge, a perennial problem of municipal wastewater treatment plants, is less severe a problem in most food and agricultural wastewater treatment processes. Detailed information regarding typical wastewater characteristics including heavy metals, can be found in Metcalf & Eddy (1991).

7.2 Sludge thickening (concentration)

Further sludge concentration is first accomplished by the use of thickening equipment that will increase the solids content to between 2–5%. Sludge thickening is achieved through one of two means: flotation or settling to the bottom by gravity or centrifugal force. The thickening operation separates water from the sludge as much as possible and is cost-effective; the cost involved in the process is well offset by the savings gained through the reduction of sludge volume, which decreases the capital and operating costs of subsequent sludge processing steps. In addition to reduction in volume, which needs to be handled

Table 7.1 Air flotation parameters for DAF

Parameter	Typical value
Air pressure, psig	40–70
Effluent recycle ratio, % of influent flow	30–150
Detention time, hours	3
Air-to-solids ratio, lb air/lb solids	0.02
Solid loading, lb/ft^2·day	10–50
Polymer addition, lb/ton dry solids	10

mechanically, sludge thickening is beneficial to the stabilization process (e.g., anaerobic digestion process) because it reduces biomass volume, tank size, and heating requirements.

The flotation process in the sludge thickening is, in principal, similar to the flotation process for pre-treatment of wastewater. Air is injected into the sludge tank under pressure from the bottom, and a large amount of air bubbles disperse into the sludge, attach themselves to the sludge solids and float them to the surface of the thickener. The layer that is formed by the floated sludge particles is removed from the surface by a skimming mechanism for further processing.

The flotation method is particularly suitable for activated sludge because of the low specific gravity of the solids, which makes it difficult for gravity-based thickening to remove or separate solids from water. The common parameters for dissolved air flotation process are shown in Table 7.1.

Gravity thickeners are essentially circular primary settling tanks, with or without mechanical thickening devices. The dilute sludge is fed into the settling tank, where solids are allowed to settle over a few days. The thickened sludge is withdrawn from the bottom of the tank and pumped to the digesters or dewatering equipment. Gravity thickening can be achieved in a separate tank or within the clarifier, if it is so designed. Thickening within the clarifier is achieved at the lowest part of the clarifier within a sludge storage zone or hopper. Within the hopper, the sludge is slowly mixed with a motorized rack to enhance the release of water. The mechanical stirrer rack is comprised of a rotating set of vertical blades or rods, which gives the stirrer an appearance resembling a picket fence, giving rise to the name, "picket fence thickener" (Figure 7.1).

Often, a secondary tank is used to supplement or replace the thickening zone of the clarifier, especially when extended thickening times are required. These units are typically designed to store the accumulated solids for at least 24 hours. Chemicals (e.g., iron and aluminum salts, polyelectrolytes) are sometimes added to aid the thickening process, in a process called sludge conditioning. Plain settling tanks can produce solids content in sludges of up to 8.0% for primary

7.2 SLUDGE THICKENING (CONCENTRATION)

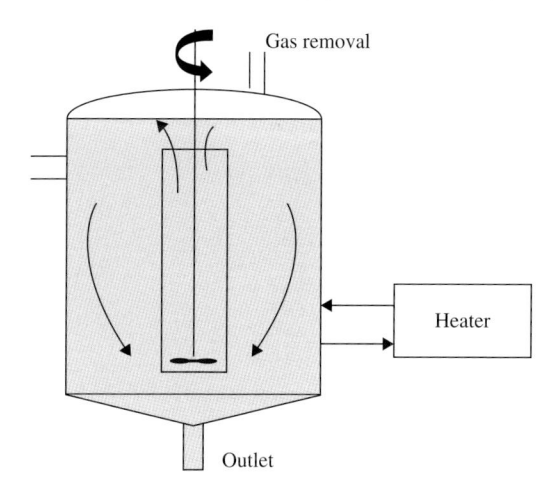

Figure 7.1 Schematic diagram of a picket fence gravity thickener

sludges and up to 2.2% for activated sludge. Activated sludge can also be concentrated by resettling in primary settling tanks.

Gravity thickener design is very similar to that of primary settling basins, and the mechanism of the process is also similar. Due to relatively higher solid content in sludge than that in wastewater, a heavy-duty scraper is often called for in sludge gravity thickening, in order to move sludge to a hopper, from which it is withdrawn and further processed. Gravity sludge thickening is often used for primary sludge thickening, while flotation thickening is frequently employed for activated sludge thickening. A gravity thickener will be designed on the basis of hydraulic surface loading and solids loading. The design principles are to be the same as those for sedimentation tanks.

The use of chemical additives (lime or polyelectrolytes) allows higher loading rates. The minimum detention time and the sludge volume divided by sludge removed per day (which represents the time sludge is held in the sludge blanket) is usually less than two days. The design parameters of gravity thickeners for different types of sludges are listed in Table 7.2.

Table 7.2 Mass loading for designing sludge thickeners

Type of sludge	Mass loading, lb/ft^2·day
Primary sludge	22
Primary and trickling filter sludge	15
Primary and waste activated sludge	6–10
Waste activated sludge	4–8

Figure 7.2 Aerial photo of Gold Bar Wastewater Treatment Plant in Edmonton, Alberta, Canada (see plate section for color version)

Centrifuges are used both in thickening and dewatering of sludge. Centrifugal thickening involves the settling of particles in watery sludge under the influence of centrifugal forces. The application of centrifuges is limited to activated sludge. The advantage of centrifugal thickening is its ability to thicken some difficult sludge. The downsides are the costs associated with power and maintenance, as well as skilled operators. Other types of sludge thickening equipment are rotary drum thickener and gravity belt thickener, but these are less common.

The main design variables to be considered in selecting a thickening process are:

- Solids concentration and volumetric flow rate of the feed system
- Chemical demand and cost if chemicals are employed
- Suspended and dissolved solids concentrations and volumetric flow rate of the clarified stream
- Solids concentration and volumetric flow rate of the thickened sludge.

Figure 7.3 Photo of a reed bed sludge treatment before planting vegetation (see plate section for color version)

Other variables that impact the selection of a thickening process are: subsequent processing steps; operation and maintenance (O/M) cost; and the reliability required for meeting successful operational requirements.

7.3 Sludge stabilization

Once the sludge is thickened, there are two options available for further treatment of the concentrated sludge. It can be dewatered to a solid content of between 30–40%, or it can undergo stabilization processes to reduce the organic materials in the sludge before going to the dewatering step. Coarse primary solids and secondary sludge (sometimes called biosolids) accumulated in a wastewater treatment process must be treated before disposal to ensure an environmentally responsible and lawful outcome. Sludge is often inadvertently contaminated with toxic organic and inorganic compounds and is nutrient-rich.

The objective of sludge stabilization is multi-facet: it reduces pathogens, eliminates odor, and reduces organic matters preventing or inhibiting future decomposition. This is relevant to sludges from many food processing wastewaters. Sludges are stabilized to prevent anaerobic breakdown occurring naturally during storage (a process termed as putrefaction) and thereby producing offensive odors. Stabilization can be done chemically or biologically, though the latter is more common and more effective.

Lime stabilization is achieved when a sufficient amount of lime is added to the sludge to alter the pH value to a high level (>11) that no microorganisms

Figure 7.4 Photo of a reed bed sludge treatment after planting vegetation (see plate section for color version)

can survive. A similar lime treatment can be applied after sludge dewatering to achieve the same objective. Since lime treatment does not destroy or alter any organic matter, an excessive amount of lime is often required to maintain the high pH value that is needed to prevent future decomposition of organic solids in the sludge.

Biological stabilization utilizes biological (in many cases, microbiological) agents to reduce organic matters in the sludge – a process often termed "digestion." There are various digestion techniques, the purpose of which are to reduce, in addition to the amount of organic matter, the number of disease-causing microorganisms present in the solids. The most common treatment options include anaerobic digestion, aerobic digestion, vermistabilization, and composting.

Anaerobic digestion is the most common and widely used sludge stabilization process – and it also helps to reduce global warming. If sludge were landfilled without stabilization, it would still break down naturally (and anaerobically, most likely). However, biogas (a mixture of 55% to 75% methane and other gases, mainly CO_2) would escape directly into the atmosphere, and CH_4 is a worse greenhouse gas than CO_2. Because of this, anaerobic digestion is considered to be a sustainable technology, and biogas is considered to be a renewable fuel if utilized.

7.3 SLUDGE STABILIZATION

7.3.1 Sludge aerobic digestion

Aerobic digestion is a process of treating the secondary sludge from biological wastewater treatment processes such as activated sludge and trickling filters, while primary sludge is better treated by anaerobic digestion (see Chapter 4). The secondary sludge is primarily comprised of insoluble solids such as a biomass of microorganisms. The objective of aerobic digestion is to degrade insoluble solids in an aerobic environment. Aerobic digesters are simply CSTRs, not much different from those used in activated sludge. Both bubbling and mechanical aerators achieve mixing of oxygen into the liquid in the tank. By optimizing the oxygen supply, the process can be significantly accelerated.

The aerobic digestion is designed to treat excessive amounts of sludge from the activated sludge process and other biological treatment processes. Early attempts to treat this type of sludges with anaerobic digestion met with little success because of its low solids content and the highly aerobic nature of the sludge. The high water content (98–99%) of this type of sludge also prevents economical dewatering by mechanical means without substantial thickening. In small communities, the high capital investment requirement associated with thickening and anaerobic digestion equipment also prohibits the use of anaerobic digestion; these communities are likely to choose aerobic digestion instead.

An aerobic digester is normally operates by continuously feeding the raw secondary sludge into the tank, punctuated with supernatant and sludge withdrawals. The aerobic digester is aerated continuously while the tank is being filled, as well as for the period immediately following that. Once aeration is stopped, the solids are allowed to settle by gravity. The supernatant is decanted and a portion of the gravity-settled sludge is withdrawn.

7.3.2 Sludge anaerobic digestion

Anaerobic digestion is a bacterial process that is carried out in the absence of oxygen. The process can either be thermophilic digestion, in which sludge is fermented in tanks at a temperature of 55°C, or mesophilic, at a temperature of around 36°C. Although it allows shorter retention time and, thus, smaller tanks, thermophilic digestion is more expensive in terms of the energy consumption required for heating the sludge.

Anaerobic digesters have been around for a long time, and they are commonly used for sewage treatment and for managing animal waste. Increasing environmental pressures on waste disposal has increased the use of digestion as a process for reducing waste volumes and generating useful byproducts. It is a fairly simple process that can greatly reduce the amount of organic matter that might otherwise end up in landfills or waste incinerators.

Almost any organic material can be processed in this manner. This includes biodegradable waste materials such as waste paper, grass clippings, leftover food, sewage, and animal waste. Alternatively, anaerobic digesters can be fed with specially grown energy crops to boost biogas production. After sorting or screening to remove inorganic or hazardous materials such as metals and plastics, the material to be processed is often shredded or minced to achieve a better reaction (ultrasound has even been used in the process to aid in the break-up of solids). Breaking the material into smaller pieces provides the bacteria with a much greater surface area, allowing them to complete the process quicker. The material is then fed into a sealed digester. In the case of dry materials, water is added.

Anaerobic digestion generates biogas with a high proportion of methane that may be used to both heat the tank and run engines or microturbines for other on-site processes. In large treatment plants, sufficient energy can be generated in this way to produce more electricity than the machines require. The methane generation is a key advantage of the anaerobic process. Its key disadvantage is the long time required for the anaerobic process (up to 30 days) and the high capital cost.

The Gold Bar Wastewater Treatment Plant in Edmonton, Alberta, Canada currently uses this process (Figure 7.2). Under laboratory conditions, it is possible directly to generate useful amounts of electricity from organic sludge using naturally occurring electrochemically active bacteria. Potentially, this technique could lead to an ecologically positive form of power generation but, in order to be effective, such a microbial fuel cell must maximize the contact area between the effluent and the bacteria-coated anode surface, which could severely hamper throughput.

7.3.3 Vermistabilization

Vermistabilization is a technology that utilizes earthworms to stabilize and dewater wastewater sludge. It provides an all-in-one approach to sludge treatment. The technology only works for sludges with sufficient organic matters and nutrients to support the earthworm population. *Eisenia foetida* has been shown to be the best earthworm species for this, due to its growth rate and reproductive responses, with temperatures ranging from 20–25°C (68–77°F).

Although vermistabilization has been used and studied in composting in many nations, its application in sludge treatment is far from certain. There are a number of critical issues that prevent vermistabilization from being a common practice, among them being that earthworms can, and do, accumulate heavy metals and other organic pollutants. Past studies have also found that some viruses, bacteria and parasites can pass through the guts of earthworms and survive. In addition, certain industrial or even municipal sludges contain or do not contain certain substances that adversely affect the growth and reproduction of earthworms.

These shortcomings of vermistabilization may not be crucial for many food processing wastes. However, if food and agricultural wastewaters mix with other industrial or municipal wastewater, or food and agricultural wastewaters contain toxic materials or heavy metals, then vermistabilization may not be applicable to the sludge originating from these wastewaters.

7.3.4 Composting

Composting is also an aerobic process for the concurrent stabilization and dewatering of sludges. It involves mixing the wastewater solids with sources of carbon, such as sawdust, straw or wood chips, to enable the biological process. In the presence of oxygen, bacteria digest both the sludge and the added carbon source and, in doing so, they produce a large amount of heat. There are three basic types of compost systems (Reed et al., 1995): windrow, static pile, and enclosed reactors:

- In a windrow system, the mixture of sludge and wood chips to be compost is placed in long rows, which are periodically turned and mixed to expose new surfaces to oxygen in the air.
- Static pile systems consist of a porous base made from wood chips or compost, in which the air is blown or drawn through a either perforated or non-perforated pipe. The wood chips and sludge is piled on top of the porous base, and screened compost covers the sludge-wood chip mixture.
- Enclosed reactors look more or less like miniature compost containers used by home gardeners. Inside the reactors, there could be static pile or windrow-type layouts; the enclosure in this is usually for odor control.

Table 7.3 provides a general guideline for designing compost systems for sludge treatment. Monitoring process parameters is essential in any composting operation, as it ensures efficient operation and quality of final products. Critical parameters such as moisture, oxygen concentration, heavy metals and organics, pathogens, pH, and temperature, need to be watched closely and continuously for successful operations and for compliance with laws and regulations.

7.4 Reed beds

Reed beds constructed in wetlands provide long-term storage and volume reduction of sludges (sludge stabilization and dewatering) to mitigate environmental

Table 7.3 Design considerations for aerobic sludge composting

Item	Design considerations
Sludge sources	Untreated sludge or digested sludge.
Amendments and bulking agents	Wood chips, sawdust, recycled compost, and straw.
Carbon : nitrogen ratio	Ranging from 25 : 1 to 35 : 1
Volatile solids	> 50%
Air requirement	> 50% remaining in all parts of compost
Moisture content	Less than 60% for static pile type; less than 65% for windrow.
pH	6 to 9
Temperature	Initially 50 to 55°C and later 55 to 66°C
Mixing and turning	Periodically, depending type of compost operations.
Heavy metals and trance organics	Do not exceed the limits for compost disposal.
Site constraints	Available area (minimum area for accommodating 30 days of composting production), access, proximity to treatment plant, climatic conditions, and availability of buffer zone.

Source: Adapted from Metcalf & Eddy (1991).

concerns. Widely used throughout Europe, Asia, and Australia, and in more than 50 locations in the United States, reed-bed technology features low construction costs and minimal day-to-day operation and maintenance costs, due to infrequent cleaning of the beds (usually a cycle of several years). The reed bed system reduces water content, minimizes sludges, and provides sufficient storage time to stabilize sludges prior to disposal.

Phragmites was first used years ago in Europe for handling iron oxide sludges. Reed beds use common reed plants (*phragnmites communis*) to dewater solids in a confined area. The beds can be any shape to accommodate existing land conditions and areas. Specially designed ponds, with under-drains covered by a sand and gravel mixture, are constructed and filled with reed plants. Modified sludge drying beds also work well and are an ideal retrofit. They already have sidewalls, layers of sand and gravel, an under-drain system that collects and carries away filtrate, and an impervious membrane liner (See Figures 7.3 and 7.4 for reed bed in use).

Solids are pumped into the reed beds and dewatering occurs through evaporation, plant transpiration, and decantation. Decanted water seeps through the bottom of the bed and through the layers of sand and gravel into the under-drains, traveling back to the wastewater treatment plant for secondary treatment. During dewatering, the solids change from liquid to "cake". Six inches of solids and water will compress to a half-inch of solid cake. The cake is left in the bed and the process is repeated.

The reeds are planted one foot on center throughout the bed. Aerobically stabilized sludge is typically applied uniformly through a grid-perforated tile. Sludges must be well stabilized, 60% volatilized or less, to be used successfully with reed beds. Optimum application rates range between 2–4% solids. While plants are young, they should be watered with plant effluent. After they are established, they can be fed heavier sludge mixtures. Loading rates (gallons per square foot per year or liters per square meter per year) depend on temperature for well-established beds.

The phragmite is one of the most widespread flowering plants in the world. It is a tough and adaptable plant, which can grow in polluted waters and find sustenance in sludge. This reed has a voracious appetite for water and is tolerant to low oxygen levels and to waterlog conditions. The reeds hold themselves in the soil through roots and rhizomes, an intricate network of underground stems. New plants, in turn, will sprout from these stems. These rapidly growing roots provide air passages through the sludge that, in turn, provide a host area for many biological communities to develop and continue to mineralize the sludge.

Reed beds perform three basic functions:

1. dewater the sludge;
2. transform it into mineral and humus-like components;
3. store the sludge for a number of years.

Dewatering is accomplished through evaporation (as in a normal sludge drying bed operation), transpiration through the plant's root stem and leaf structure, and filtration through the bed's sand and gravel layers and the plant's root system. The leachate is channeled back to the treatment plant through the under-drain.

The plants should be harvested annually to prevent drainage backup. The vegetation can be composted or burned.

Researchers recommend the installation of multiple beds to handle emergencies and downtime due to cleaning. Beds may be out of service for up to a year while rootstalks grow new tops after cleaning. The top level of sand and material removed during clean-out is similar in pathogen content to composted sludge and can be used in the same way. Many beds have gone eight to ten years without having to be cleaned out.

Sludge reed beds are a significant improvement over existing drying beds. Sludge can be dewatered and converted into biomass and low-grade compost without the addition of chemicals or expenditure of energy. They have a lengthy turnover time and are capable of reducing sludge volumes by up to 95% over time.

7.5 Conditioning of sludge

The aim of conditioning the sludge is to improve its dewatering characteristics. The dewaterability of sludges varies; in some types, such as activated sludge, it is difficult – something that is mainly attributed to the presence of extracellular polymer (ECP). ECP is present in varying quantities in sewage sludge, occurring either as a highly hydrated capsule surrounding the bacterial cell wall, or loose in solution as slime polymers. ECP is thought to aid the survival of the bacterial cell by preventing desiccation and acting as an ion-exchange resin, controlling the ionic movement from solution into the cell. Polysaccharide, protein and DNA, which entrap the water and cause high viscosity, are the main components of ECP, but humic-like substances, lipids and heteropolymers such as glycoproteins are also present.

Wang *et al.* (2004) noted that surface properties, like the concentration of ECP, were related to zeta-potential measurements with particle electrophoresis as well as to water contact angle measurements on filter cakes prior to and after the oxidative conditioning. The sludges with high zeta potentials and low contact angles were those sludges with high amounts of ECP. These sludges, which have a high surface charge density as well as a high hydrophilicity (low contact angle) prevent efficient flocculation.

The experimental results of Neyens *et al.* (2004) indicate that peroxidation of sludge enhances the flocculation and dewaterability. The responsible mechanism is not fully understood, but the oxidative conditioning might be based on partial oxidation and rearrangement of the surface components (extracellular polymers) of the sludge flocs. The effects of temperature, hydrogen peroxide concentration, pH, presence of Fe^{2+} and reaction time on the dewaterability of the sludges were tested by Neyens *et al* (2002, 2004). They found that peroxidation, among all advanced sludge treatment processes, gave the best results with respect to improving sludge dewaterability and product quality of the residual filter cake.

The general approaches to conditioning sludge are employment of chemicals (similar to coagulants used in primary sedimentation) and heat treatment to improve the dewaterability of sludge. Other less common methods, such as freezing, irradiation, solvent extraction and, as mentioned previously, advanced oxidation, are still in the laboratory stages.

Chemical conditioning can reduce 90–99% moisture content of an incoming sludge down to 65–85%. As described previously, chemicals such as coagulants (e.g., alum, lime, iron salts, and polymers) disrupt the structure and composition of the sludge by forming aggregates that are more compact in structure and hold less water. The addition of chemicals may contribute to increasing the solid content of the sludge as much as up to 30%. The dosage and types of chemicals depends on several factors, but the properties of the sludge are by far the most crucial factors in determining which and how much chemical needs to add to the sludge. The source of the sludge, solids concentration, the age of the sludge, pH, and alkalinity are important properties that affect the selection and dosage of chemicals added to the sludge prior to mechanical dewatering operations.

It has been generally observed that the difficult-to-dewater sludges require larger quantities of chemicals. The more biologically processed the sludge is, the larger the dose of chemicals that is required. In this vein, aerobically digested sludge that is treated with the most efficient biological process requires the highest dose of chemical conditioners, while untreated raw primary sludge needs the least.

Heat treatment can serve both as a stabilization process and as a conditioning process. Heat can facilitate aggregation of particles, break down the gel structure of the sludge, and reduce water affinity to the sludge particles. Heat can also sterilize or inhibit pathogens, bacteria, and enzymes. Heat treatment of sludge, if done properly and followed with vacuum filtration or belt presses, seldom causes putrefaction. The other benefit of the heat-processed sludge is that it has a heating value of 12,000 to 13,000 But/lb (28 to 30 kJ/g) of volatile solids (volatile solids are part of the dried sludge that undergoes complete combustion in a muffle furnace at 550°C), and this can be valuable in sludge incineration.

The disadvantages of heat treatment of sludge are also noticeable. They include: the costs, both capital and operating, which are high due to mechanical complexity; operations of heat treatment require skilled operators and elaborate operating procedures; significant odor and gases emit from the process that need to be managed; and scale formation in heat exchangers is extensive and difficult to prevent or clean up.

7.6 Dewatering

Dewatering of sludge is a physical unit operation to reduce the moisture of the sludge in preparation for subsequent further treatment processes. As the agricultural use of the sludge and landfilling are increasingly restricted, both in the United States and elsewhere, drying and incineration are widely implemented. As a result, the costs related to the treatment of sludge have risen considerably and commonly represent 35–50% of the total operating costs of the wastewater treatment. Reducing the amount of sludge produced and improving the dewaterability are thus of paramount importance. This objective of sludge reduction has stressed the importance of using extended aeration biology, using biological phosphorus-removal (instead of chemical precipitation), using sludge digesters, etc. Further reduction and improvement of the dewaterability will require advanced sludge treatment technologies, such as Fenton reaction and peroxide oxidation (Neyens *et al*, 2002).

A number of techniques of dewatering are used for removing moisture. Some rely on natural forces such as evaporation and percolation, while the others are assisted mechanically and thermally during dewatering. The objective of dewatering is moisture content of 60–80%, depending on the disposal method. Dewatering of the sludge is currently achieved through the use of mechanical dewatering and thermal dehydration devices. The most widely used mechanical dewatering device is the filter press, while other technologies employed include vacuum filters, centrifuges and belt presses. Mechanical dewatering will produce a sludge with an approximately 10–60 percent solids content (e.g., using common centrifuges or belt presses, only 20–25% dry solids can be obtained). Sludge dryers are used to further remove moisture from the sludge and are capable of producing a material with 90% solids content. Because all dewatering devices are dependent upon proper sludge conditioning, a carefully designed chemical feed system should be included as part of the dewatering facility.

7.6.1 Belt press filtration

Belt filter presses employ single or double moving belts to continuously dewater sludges through one or more stages of dewatering. All belt press filtration processes include three basic operational stages: chemical conditioning of the feed sludge; gravity drainage to a non-fluid consistency; and shear and compression dewatering of the drained sludge. When dewatering a 50 : 50 mixture of anaerobically digested primary and waste activated sludge, a belt filter press will typically produce a cake solids concentration in the 18–23% range.

Figure 7.5 depicts a simple belt press and shows the location of the three stages. The dewatering process is made effective by the use of two endless belts of synthetic fiber. The belts pass around a system of rollers at constant speed and perform the functions of conveying, draining and compressing. Many belt presses also use an initial belt for gravity drainage, in addition to the two belts in the pressure zone.

7.6.2 Centrifugation

Centrifugal dewatering of sludge is a process that uses the force developed by fast rotation of a cylindrical drum or bowl to separate the sludge solids from the liquid. In the basic process, when sludge slurry is introduced to the centrifuge, it is forced against the bowl's interior walls, forming a pool of liquid. Density differences cause the sludge solids and the liquid to separate into two distinct

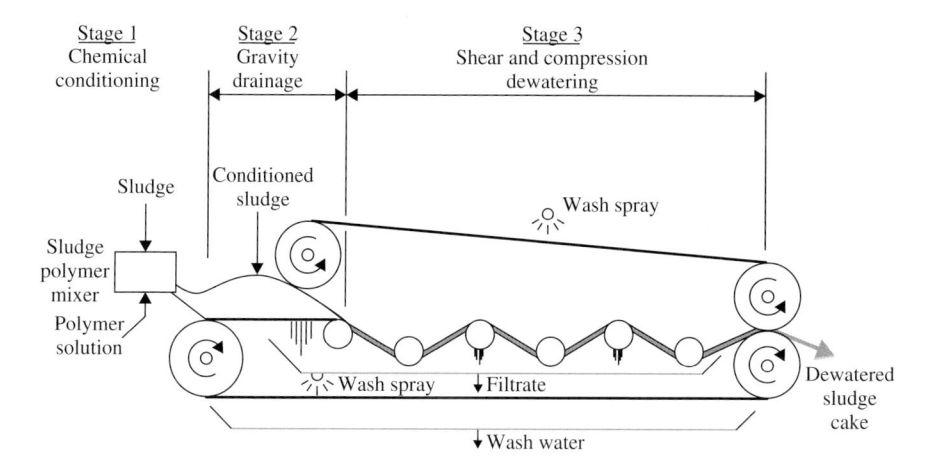

Figure 7.5 A schematic diagram of belt filter for sludge dewatering

layers. The sludge solids "cake" and the liquid "centrate" are then separately discharged from the unit. The two types of centrifuges used for sludge dewatering are basket and solid bowl, and both operate on these basic principles. They are differentiated by the method of sludge feed, magnitude of applied centrifugal force, method of solids and liquid discharge, cost and performance.

Basket centrifuge

The imperforate basket centrifuge is a semi-continuous feeding and solids discharging unit that rotates about a vertical axis. A schematic diagram of a basket centrifuge in the sludge feed and sludge plowing cycles is shown in Figure 7.6. Sludge is fed into the bottom of the basket and the sludge solids form a cake on the bowl walls as the unit rotates. The liquid (centrate) is displaced over a baffle or weir at the top of the unit. Sludge feed is either continued for a preset time or until the suspended solids in the centrate reach a preset concentration. The ability to be used either for thickening or dewatering is an advantage of the basket centrifuge. A basket centrifuge will typically dewater a 50 : 50 ratio blend of anaerobically digested primary sludge and waste activated sludge to 10–15 percent solids.

Bowl centrifuge

Solid bowl centrifuges are a type of centrifuge which has a rotating bowl into which sludge is fed at a constant flow rate. Chemicals or polymers can be added to aid the dewatering. The centrate obtained from the separation usually contains fine solids and is returned to the wastewater treatment systems. The cake

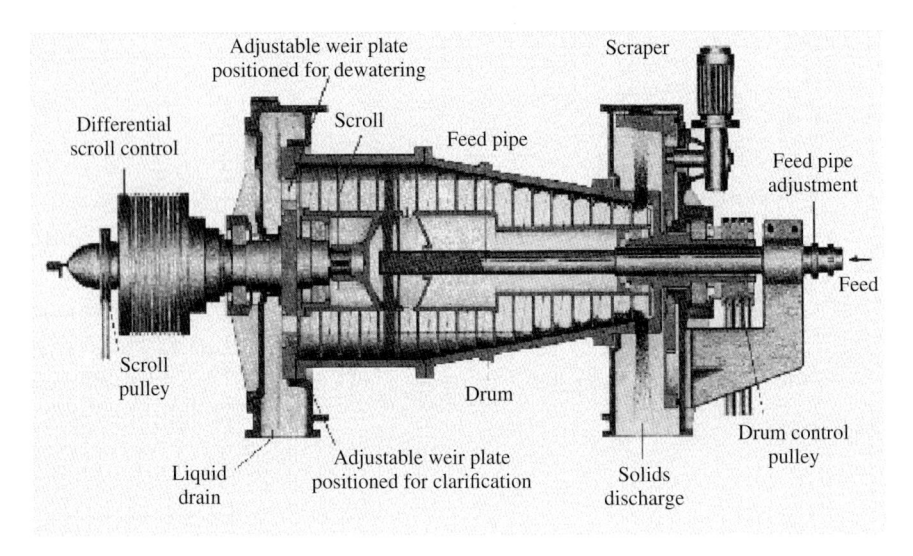

Figure 7.6 A schematic illustration of centrifuge for dewatering of sludge (see plate section for color version)

is 70–80% moisture and is discharged from the bowl by a screw feeder into a hopper or a convey belt. Sludge cakes with greater than 25% solids are desirable for disposal by incineration or landfill.

7.6.3 Filter press

The filter press is an intermittent dewatering process. A filter comprises a set of vertical, juxtaposed recessed plates, pressed against each other by hydraulic jacks at one end of the set. The joint face of each filtering plate must be sufficiently strong to withstand the chamber internal pressure developed by the sludge pumping system.

This vertical plate layout forms watertight filtration chambers, allowing easy mechanization for the discharge of cakes. Filter cloths, tightly meshed, are applied to the two grooved surfaces in these plates. Orifices feed the sludge to be filtered under pressure in the filtration chamber. They are usually placed in the center of the plates, allowing a proper distribution of flow, a suitable pressure and better drainage of sludge within the chamber. Solid sludge gradually accumulates in the filtration chamber until the final compacted cake is formed. The filtrate is collected at the back of the filtration support and carried away by internal ducts (Figure 7.7).

The filter press has a number of advantages over other filtration equipment, such as vacuum filters and centrifuges. Filter presses can operate well at variable

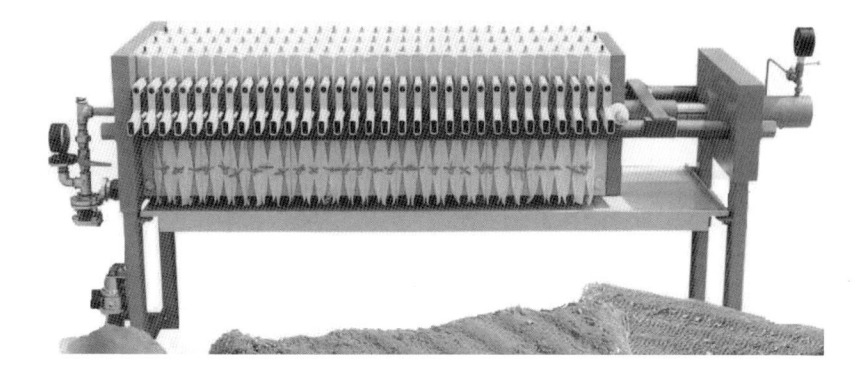

Figure 7.7 A filter press for sludge dewatering (see plate section for color version)

or low feed conditions. They can also produce a relatively dry cake because of the high-pressure differential that they can exert on the sludge. A typical filter press operating at 790.8 KN/m^2 (100 psig) will produce sludge with a solids content of 25–60% solids, depending on the chemicals used for precipitation. By comparison, a basket centrifuge produces sludge with a solids content of 10–25% and a vacuum filter produces sludge with 15–40% solids.

The disadvantages of the filter press include its batch operating cycle, the labor associated with removing the cakes from the press, and the downtime associated with finding and replacing worn or damaged filter cloths. The original filter press design consisted of alternating plates and frames, and these types of units were referred to as the plate-and-frame filter press. The new and improved design is the recessed plate filter press. Here, the plates (usually constructed of polypropylene) are recessed on each side to form cavities, and they are covered with a filter cloth.

The two types of presses work in basically the same manner. At the start of a cycle, a hydraulic pump clamps the plates tightly together and a feed pump forces dilute sludge slurry into the cavities of the plates. The liquid (filtrate) escapes through the filter cloth and grooves that are molded into the plates, and is transported by the pressure of the feed pump to a discharge port. The solids are retained by the cloth and remain in the cavities. This process continues until the cavities are packed with sludge solids. The hydraulic pressure is then released and the plates are separated. The sludge solids or cake is loosened from the cavities and falls into a hopper or drum.

Modern recessed plate filter presses may be equipped with the following design enhancements:

- Lightweight polypropylene plates that exhibit good chemical resistance and provide a long service life.

- Gasketed plates that reduce leakage during the filtration cycle. These replace non-gasketed types, where the filter cloth extends beyond the plate to form the seal between plates.

- An air blow-down manifold that is employed at the end of the filtration cycle to drain remaining liquid in the system, thereby improving sludge dryness and aiding in the release of the cake.

- Microprocessor control, which permits unattended operation throughout the filtration cycle. It is capable of adjusting the feed pressure automatically and deactivating the pump whenever hydraulic pressure falls below preset limits.

- Manual, semi-automatic or automatic plate shifters that are used to separate the plates prior to releasing the sludge cake.

7.6.4 Sludge freezing

Freezing and then thawing sludge will transform the structure of the sludge to make it amenable to draining of interstitial water trapped within it. The expansion of sludge structures due to formation of ice crystals during freezing apparently also helps to loosen the binding of water molecules to the sludge solids. Freezing and thawing processes of waste activated sludge have been investigated extensively (e.g., Reed *et al.*, 1986), and it has been found that surface water can be expelled by freezing. When water in sludge begins to freeze, it creates a thin upper layer that sends needles into the sludge. As ice growth continues, then the smaller the solid particles are, the faster advancing ice front moves them. Some large particles (perhaps greater than 100 µm) cannot be pushed in front of the ice and are trapped within the frozen mass without being moved. In time, the ice crystals dehydrate the captured sludge flocs, pushing the particles into more compact aggregates.

Freezing/thawing process can effectively take place in a "freezing bed" as proposed by Martel (1989) (see Figure 7.8). The freezing bed operates somewhat like a sludge drying bed, except that the sludge is applied in thin layers during the winter months and allowed to freeze. During warmer weather, the sludge thaws and the water drains out, leaving a dry residue (Figure 7.9).

The freezing/thawing treatment of activated sludge is not economically feasible unless natural freezing and thawing are used. In the United States, the freezing and thawing method for dewatering of sludge is not suitable for eastern regions south of the Mason-Dixon line, California, most of Arizona and New Mexico, and parts of the northwest coastal states.

A simple equation that can be used for preliminary assessment of feasibility of the freezing and thawing method for sludge dewatering correlates the total depth

7.6 DEWATERING

Figure 7.8 A freezing-thawing sludge bed (see plate section for color version)

Figure 7.9 A sample of freezing-and-thawing treated sludge from anaerobic digester (see plate section for color version)

of sludge that could be frozen if applied in 8 cm increments with the maximum depth of frost penetration (Reed *et al.*, 1995):

$$\Sigma Y = 1.76 \, (F_p) - 101 \text{ (metric units)} \tag{7.1}$$
$$= 1.76 \, (F_p) - 40 \text{ (imperial units)}$$

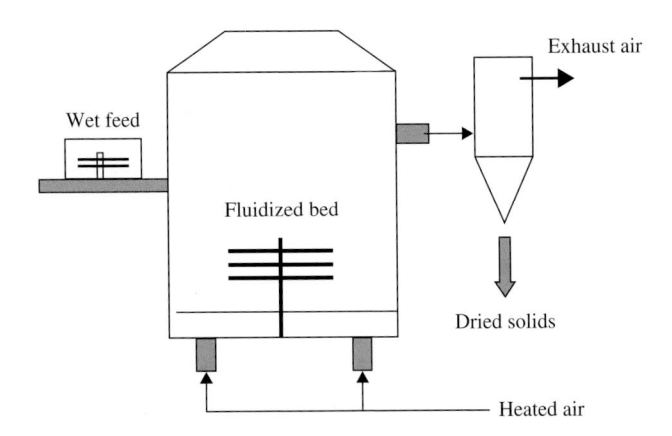

Figure 7.10 Schematic diagram of flash dryer for sludge dewatering

where:

ΣY = the total depth of sludge that can be frozen in 8 cm (or 3 in) layers during the warmest design year, cm or in

F_p = the maximum depth of frost penetration, cm or in.

7.6.5 Sludge drying

Sludge drying is achieved through vaporization of water in sludge. There are two categories of drying sludge. One is a drying bed type; the other is a mechanical device type that requires auxiliary heat to increase vaporization of moisture in sludge. Drying by natural means is only possible during a long period of time. Mechanical processes are faster and smaller, but also more cost-intensive. Sludge drying beds are used for small communities for dewatering sludge from wastewater treatment. The beds are basically a constructed storage area for holding sludge, and dewatering (drying) is achieved by draining the sludge by gravity and vaporization of moisture from the sludge exposed to the atmosphere. Some designs employ air to create a vacuum under the draining system of the bed. After the drying process, the dried sludge is destined for landfill. The principal advantage of drying beds is low cost and low maintenance. The disadvantage is the cost of removing the sludge and replacing draining bedding (sand). Thus, drying bed type dewatering operations are suitable for large communities with populations over 20,000.

Mechanical processes of heating sludge include flash dryer (Figure 7.10), spray dryer (Figure 7.11), rotary dryer (Figure 7.12), multiple-hearth incinerator (discussed in Section 7.8), and multiple-effect evaporator (Figure 7.13).

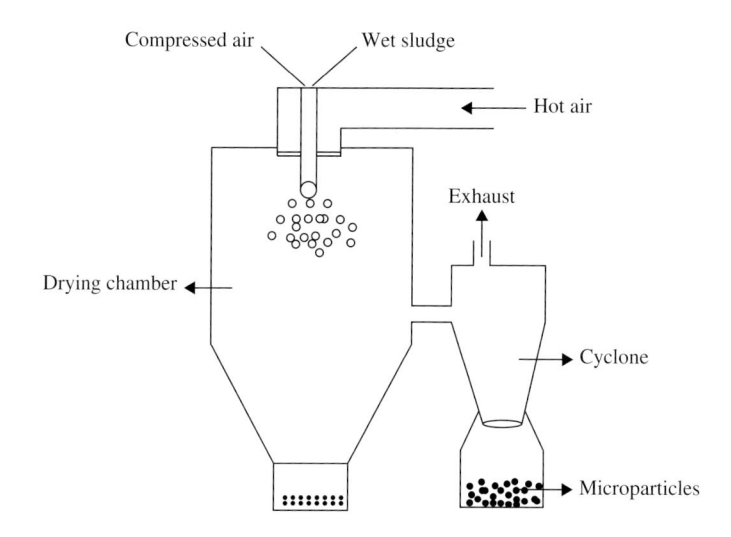

Figure 7.11 Schematic diagram of spray dryer for sludge dewatering

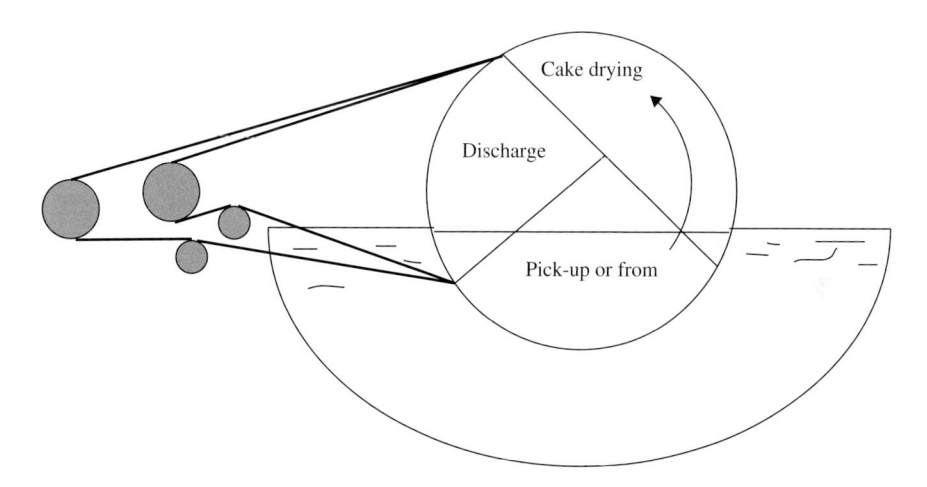

Figure 7.12 Schematic diagram of rotary dryer for sludge dewatering

Flash dryers involve pulverizing sludge in a cage mill or in the presence of hot gases. The process is based on exposing fine sludge particles to turbulent hot gases long enough to attain at least 90% solids content. The dried sludge may be used as soil conditioner, or it may be incinerated.

Spray dryers typically use centrifugal force to atomize liquid sludge into a spray that is directed into a drying chamber, where it contacts with hot air that rapidly dries the sludge mist into powder. The flow directions of hot air and the sludge streams are either concurrent or counter-current.

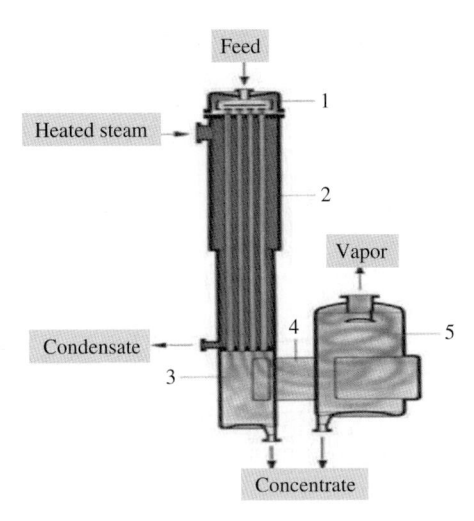

Figure 7.13 Schematic diagram of multiple-effect evaporator for sludge dewatering (see plate section for color version)

Rotary dryers function as horizontal cylindrical kilns. The drum rotates and may have plows or louvers that mix the sludge mechanically as the drum turns. There are many different rotary kiln designs, utilizing either direct or indirect heating systems. Direct heating designs maintain contact between the sewage sludge and the hot gases. Indirect heating separates the two with steel shells.

Multiple-effect evaporators in sludge drying operations use the proprietary multi-effect Carver-Greenfield process, in which dewatered sludge is mixed with oil. This mixture, which can be pumped easily, is pumped through a series of evaporators (multi-effect evaporation system) that selectively remove the water in the sludge, which has a lower boiling point than the oil. The oil maintains the mixture in a liquid state even when virtually all the water has been removed. The product of this process, an oil and dry sludge mixture, is put through a centrifuge to separate the dry sewage sludge solids from the oil. The recovered oil can be reused in the process.

7.7 Land applications and surface disposal

Land application of sludge is defined as the beneficial use of the sludge at agronomic rates, while all other land displacements are considered to be surface disposal of sludge. The detailed regulation regarding land application and surface disposal of sludge from wastewater treatment plants are discussed in 1993 EPA 40 CFR, Part 503. It stipulates that landfilling of sludge is considered as "beneficial use" only when such disposal includes methane gas recovery for fuel. However, methane operations are relatively rare. Alternative beneficial uses are receiving greater attention, because of a decline in available landfill space and an interest in conserving nutrients and utilizing soil conditioning properties and other recoverable qualities of sludge. Thus, land application for soil conditioning and fertilization is the primary beneficial use of sludges from food and agricultural wastewater treatment plants.

Sludges may be disposed of by liquid injection to land or by disposal in a landfill. There are concerns about sludge incineration because of air pollutants in the emissions, along with the high cost of supplemental fuel, making this a less attractive and less commonly constructed means of sludge treatment and disposal. There is no process that completely eliminates the requirements for disposal of sludges (for example, incineration produces ash, which needs to be properly disposed of).

Similar to land application of wastewater, land application of sludge is also affected greatly by characteristics of the waste, including organic contents (volatile solids), nutrients, pathogens, metals, and toxic organics, and soil characteristics. Metcalf & Eddy (1991) and Reed *et al.* (1995) detail the concise procedures regarding site selection, process design, and land application and surface disposal methods. The reader is encouraged to explore all aspects of sludge management issues in these references.

7.8 Incineration

Incineration, or complete combustion, is a rapid exothermic oxidation of combustible elements in fuels. The use of incineration for sludge disposal is the result of tightening limits for land disposal and/or sea disposal of sludge by regulatory agencies. The use of combustion to reduce wastewater sledges into inert ash is an effective, but costly, process unless there is a cheap fuel resource available. The ash from incineration operations is usually disposed of in a landfill.

The fuel requirement for incinerating sludges depends on two parameters: the amount of water in the sludges, and the fuel value of the contents of the sludges. For example, FOG scum from preliminary treatment or primary treatment processes of food and agricultural wastewater can be readily burned in the incinerator; this reduces the fuel requirement and lowers the cost of the operation. Raw primary and undigested secondary sludges will have fuel values ranging from about 14,000–28,000 kJ/kg (6,000–12,000 Btu/lb) dry solids (Metcalf & Eddy, 1991). If sludges can be dewatered to about 25% solids, incineration will be self-sustaining.

Oxygen requirements for incineration of sludge may be determined from knowledge of its constituents, assuming carbon and hydrogen are oxidized to their ultimate end products:

$$C_aO_bH_cN_d + (a + 0.25c - 0.5b)O_2 \rightarrow a\,CO_2 + 0.5cH_2O + 0.5dN_2 \qquad (7.2)$$

The theoretical quantity of air will be 4.23 times the calculated quantity from Equation (7.2) (O_2 is about 23% of air by weight). Excessive amounts of air (about 50% of the theoretical quantity) will be need to be used to ensure the complete combustion of the sludge.

The heat requirement for incineration operations is:

$$Q = \Sigma C_p m\,(T_2 - T_1) + m_w \lambda \qquad (7.3)$$

where:
C_p = specific heat for each constituent of substances in ash and in flue gases
m = mass of each substance
T_1, T_2 = initial and final temperatures
m_w = mass of water in sludge
λ = latent heat of evaporation of water per unit mass

It is obvious that any reduction in moisture in sludge will lower the fuel requirements. Thus, moisture content determines whether additional fuel will be required in incinerating a particular sludge.

There are two popular incinerators used for sludge incineration: multiple-hearth type and fluidized bed type.

The first multiple-hearth incinerator for sludge incineration was built in 1935 in Dearborn, MI. From that time on through the late 1960s, the multiple-hearth type was the choice for sludge incineration. At present, there are still some 150–175 multiple-hearth incinerators in operation in North America.

Multiple-hearth incinerator is a vertical cylindrical refractory lined steel shell furnace. It consists of 6–12 horizontal hearths and a rotating center shaft with rabble arms. Cooling air is introduced into the shaft, which extends above the hearths. The sludge enters the top hearth and flows downward, while combustion air flows from the bottom to the top. The rabble arms are shaped to sweep the sludge in a spiral motion, alternating in direction from the outside in, to the inside out, between hearths. The effect of the rabble motion is to break up solid material and allow better surface contact with heat and oxygen. Depending on the shaft speed and on the number of hearths, the retention time of the sludge in the incinerator ranges from 0.5–3 hours.

7.8 INCINERATION

Ambient air is first pumped through the central shaft and its associated rabble arms. A large portion of this air is then taken from the top of the shaft and recirculated back to the lowermost hearth as preheated combustion air and mixed with additional ambient combustion air. The temperature of the mixed air is limited as the lower hearths serve as an ash-cooling zone. Sludge burns in the center hearths, where it is hottest, and releases heat and combustion gas. The combustion gas flows upward through the drop holes in the hearths, counter-current to the flow of the sludge, before being exhausted from the top hearth. The flue gases rising through the furnace are cooled in the upper hearths by the evaporation of sludge moisture, which degrades or stops combustion on the top hearths. In this drying zone, some volatiles are released from the sludge and exit the furnace without exposure to the full combustion temperatures.

The feed sludge must contain more than 15% solid, due to the limitations of the evaporating capacity of the incinerator. Average loading rates range from 25–75 kg/m^2·h (5–15 lb/ft^2·h). Auxiliary fuel is often required when the feed sludge has a solids content ranging from 15–30% (Metcalf & Eddy, 1991). Additional ash handling, such as wet or dry scrubbing, is needed to minimize the air pollution.

During the 1970s, fluid bed incinerators became the preferred choice for incinerating sludge, mainly due to tighter emission regulations and to the increasing cost of the auxiliary fuel. The advantages of the fluid bed are lower emission, reduced auxiliary fuel use, and reduced operating and maintenance costs.

The fluidized bed incinerator is a vertical, cylindrically shaped refractory-lined steel shell that contains a sand bed called the combustion zone and a refractory arch containing alloy "tuyeres" or nozzles that allow hot air to be distributed homogeneously throughout the bed in order to produce and sustain combustion (Figure 7.14). The air from the refractory distributor causes the bed of sand to fluidize to a height of 5 ft. Sludge and auxiliary fuel, if required, are fed into the fluidizing sand bed through lateral feed ports. The intensive mixing of the solid and combustion air in the fluidized state yields a high heat transfer rate, resulting in rapid combustion of the sludge fed into the furnace.

The lower section area below the refractory arch distributor of the furnace is called the windbox, and this acts as a plenum in which the air is received. The refractory arch distributor and the refractory lined windbox are designed to allow combustion air to be preheated up to 676.7°C.

The section above the bed is called the freeboard or disengagement zone. It is typically 15 ft (4.6 meters) high and usually is expanded laterally along its height to maximize residence time and to reduce sand usage. The freeboard typically provides 6–7 seconds of gas residence time, which completes the combustion of any volatile hydrocarbons escaping from the bed. The freeboard thus acts as

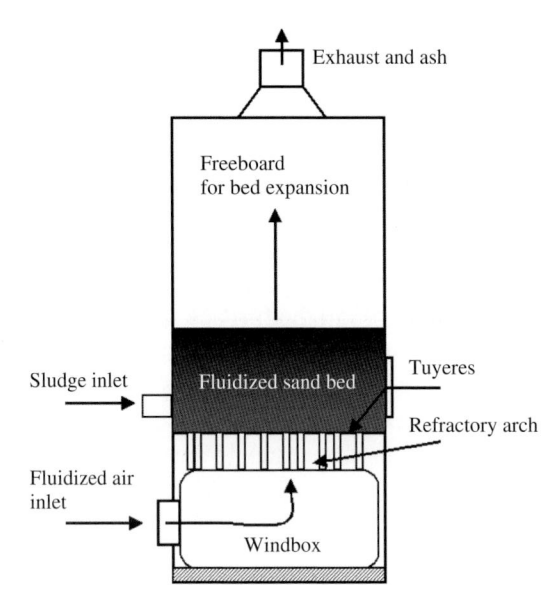

Figure 7.14 Schematic diagram of incinerator for sludge dewatering

an integral afterburner, and it normally operates at 50–100°F (10°C–37.8°C) higher than the bed, ensuring complete combustion of the volatiles.

Combustion exhaust and ash leave the bed and are transported through the freeboard area to the gas outlet through the top of the furnace. Because of fine size of the ash, no ash exits from the bottom of the incinerator. The entrained ash is scrubbed with a Venturi scrubber as part of an air pollution control system.

7.9 Further reading

Metcalf & Eddy. (1991). *Wastewater Engineering, Treatment, Disposal, and Reuse*. 3rd edition. McGraw-Hill, New York, NY.

Turovskiy, I.S. & Mathai, P.K. (2006). *Wastewater Sludge Processing*. 354pp. Wiley, New York, NY.

7.10 References

Davison, L., Headley, T. & Pratt, K. (2005). Aspects of design, structure, performance and operation of reed beds – eight years experience in north eastern New South Wales, Australia. *Water Science and Technology* **51**, 129–138.

Martel, J.C. (1989). Development and design of sludge freezing beds. *Journal of Environmental Engineering* **115**, 799–808.

Martel, J.C. (2001). Design of Freezing Bed for Sludge Dewatering at McMurdo, Alaska. ERDC/CRREL TR-01-3. US Army Corps of Engineers, Engineering Research and Development Center, Hanover, NH.

Metcalf & Eddy (1991). *Wastewater Engineering, Treatment, Disposal, and Reuse*. 3rd edition. McGraw-Hill, New York, NY.

Reed, S., Bouzoun, J. & Medding, W. (1986). A rational method for sludge dewatering via freezing. *Journal of Water Pollution Control Federation* **58**, 911–916.

Reed, S.C., Crites, R.W. & Middlebrooks, E.J. (1995). *Natural Systems for Waste Management and Treatment*. Second edition. McGraw-Hill, Inc., New York, NY.

Neyens, E., Baeyens, J., Weemaes, M. & De heyder, B. (2002). Advanced biosolids treatment using H_2O_2-oxidation. *Environmental Engineering Science* **19**, 27–35.

Neyens, E., Baeyens, J., De heyder, B. & Weemaes, M. (2004). The potential of advanced treatment methods for sewage sludge. *Management of Environmental Quality: An International Journal* **15**, 9–16.

US EPA. (1987). *Design Manual: Dewatering Municipal Wastewater Sludge*. EPA 625/1-87-014. Office of Research and Development, US EPA, Cincinnati, OH.

Wang, H., Li, X. & Zhao, C. (2004). Surface properties of activated sludge and their effects on settleability and dewaterability. *Journal of Tsinghua University (Science And Technology)* **44**, 766–769.

8

Recoverable products and energy from food and agricultural wastewaters

8.1 Introduction

The definition of "waste" according to the Merriam-Webster dictionary is "an unwanted by-product of a manufacturing process and refuse from places of human or animal habitation". However, there is another meaning of the word according to the same dictionary, which is "loss". In essence, wastes are substances whose values have yet to be determined; and, if they are not recovered or utilized, they can turn out to be a loss to society. This is a not widely shared view; throughout the developed world, food and agricultural wastewater is a scourge that needs to be kept out of sight and, hopefully, made to disappear. As a result of the prevalence of this attitude, many valuable substances present in streams or generated in wastewater treatment plants are lost.

Most wastewater management approaches are methods of concentration, conversion, and/or relocation of wastes, such as physicochemical and biological treatment, incineration, or land/sea disposal. This is in stark contrast with the early days of human development, when agricultural and food wastewaters were fully utilized as fertilizers and animal feeds. Even today, in some much less developed nations, there are still practices of utilizing food and agricultural wastewater for a variety of applications. It seems, though, that as human beings grow more affluent, their attitude towards anything associated with food and agricultural wastewaters becomes less positive.

The management and treatment of food agricultural and food wastewaters incurs costs, and this is more than an annoying nuisance for food processors and other agribusiness sectors; as environmental regulations and laws tighten, costs will undoubtedly go up. This is certainly not good news for food and agriculture

---

Food and Agricultural Wastewater Utilization and Treatment, Second Edition. Sean X. Liu.
© 2014 John Wiley & Sons, Ltd. Published 2014 by John Wiley & Sons, Ltd.

industries, whose profit margins are thin and subject to volatile international agriculture commodity price swings.

The best remedy for reducing the costs of treating and disposing food and agricultural wastewater is to reduce the amount of water used in the food and agricultural processing operations, and to recycle spent water for reuse. Water consumption in the food industry is enormous, and any reduction in its use will ultimately ease the shortage in many parts of the world and lessen environmental degradation.

Of course, certain use of water in food and agricultural processing is inevitable as the hygienic requirement for anything that handles foods to be clean and safe. Table 8.1 provides a glimpse of the amount and characteristics of food and agricultural wastewater generated in the US state of Georgia. Imagine what is like for the food and agricultural wastewaters generated throughout the whole world!

Table 8.1 Typical characteristics, estimated volume, and estimated organic loading of wastewater generated by the food processing industry in the state of Georgia in the USA

Industry sector	Estimated wastewater volume, Mgal/year	Typical characteristics	Estimated organic loading, tons/year BOD
Meat and poultry products	10,730	1,800 mg/L BOD 1,600 mg/L TSS 1,600 mg/L FOG	80,600
Dairy products	500	2,300 g/L BOD 1,500 mg/L TSS 700 mg/L FOG	14,900
Canned, frozen, and preserved fruits and vegetables	2,080	500 mg/L BOD 100 mg/L TSS	4,300
Crain and grain mill products	130	700 mg/L BOD 1,000 mg/L TSS	300
Bakery products	530	2,000 mg/L BOD	4,400
Sugar and confectionery products	140	500 mg/L BOD	300
Fats and oils	350	4,100 mg/L BOD 500 mg/L FOG	7,000
Beverages	3,660	8,500 mg/L BOD	91,000
Miscellaneous food preparations and kindred products	700	6,000 mg/L BOD 3,000 mg/L TSS	5,600

BOD: biochemical oxygen demand. TSS: total suspended solids. FOG: fats, oils, greases.
Source: Magbunua (2000). Reproduced with permission of University of Georgia, College of Engineering, Otreach Service.

There is, however, a forward-looking way of viewing wastewater as commodity whose value in our society has yet to be unlocked. From this perspective, food and agricultural wastewater is nothing but a by-product of food and agricultural processing that should be explored carefully for its possible value-added recovery and recycle.

The food industry as a whole is no stranger to this notion of recovering values from "wastes" – it has been a pioneer in using food wastes for animal foods and other non-feed uses, such as pectin recovery from apple pomace and edible oil from grape seeds (now nutraceuticals from the same "waste"). Water is routinely recovered in some food processing operations, and the need has increased as sources of fresh water perilously decline in many parts of the world. Furthermore, many sources of food and agricultural wastewater contain substantial amounts of proteins, lipids, polysaccharides, and flavoring compounds. Their recovery will undoubtedly reduce the organic loadings of wastewater treatment plants and also could offer financial returns that can be used to offset the cost of wastewater treatment and management.

However, it must be emphasized that waste utilization in food and agricultural processing operations can only be feasible if the additional cost of processing and recovering products from wastewaters is lower than the alternative (e.g., *in situ* wastewater treatment facility or paying service fees for discharging into a municipal sewage system). It is also important to remember that the utilization of solid wastes in food and agricultural wastewaters depends largely on whether the usable fraction of the wastes can be separated economically from the wastewater streams of processing plants. There is an environmental issue that is bound to be important factor in weighing the pros and cons of recovery of useful substances from wastewater versus those of treating wastewater using typical unit operations, noting that a greenhouse gas, in the form of methane (a far worse greenhouse gas than carbon dioxide), is produced in wastewater treatment operations. Unless it is recovered in the wastewater treatment plant loop, the methane produced by anaerobic reactions in a typical food and agricultural wastewater treatment is most likely to be discharged into atmosphere.

There are almost as many ways of recovering valuable materials as there are food materials in the waste. Recovered food materials from wastewater pre-treatment can be used as animal feed, or fermented to produce ethanol, or even used to make hydrogen gas (microbially produced with glucose and sucrose in the absence of oxygen). Anaerobic digestion of scum, food debris, and primary and secondary sludges can produce biogas (methane is its main useful component) as fuel. Table 8.2 is a list of the potential beneficial uses of food processing wastewaters in various sectors of the food processing industry.

Even sludges from wastewater treatment plants can be utilized beyond traditional agricultural applications. New applications of sludge products as heating

Table 8.2 Potential products from wastewaters generated from different food processing sectors of the food industry

Sector	Waste residuals in water	Potential applications
Meat, poultry, and fishery	Offal, blood, soluble proteins, DAF sludge	Animal feed, protein isolates, hormones, enzymes, savory compounds, vitamins, glue, gelatin, fish oils, and biodiesel
Bakery and grain processing	Spent brewer's yeast, starch, and waste grains	Animal feed, lactic acid fermentation ingredients, fermentation feedstock, paper and ethanol
Fruit and vegetable	Trimmings, fruit pomace, and flavors	Feed ingredients in lactic acid fermentation, animal feed, flavors, and biofuels
Nut and oilseed	Oil, hulls, meal	Fermentation feedstock, biofuels, and plastic filler (nutshell)
Dairy	Whey, lactose	Food ingredient, animal feed, fermentation feedstock for specialty chemicals, and biofuel
Beverage (non-alcohol)	Beverage spills, sugars	Fermentation feedstock for specialty chemicals
Beverage (alcohol)	Beverage spills, wine grape residuals, brewery grain residuals	Biofuels and specialty chemicals (e.g., tartarate)

fuels, ingredients for cement production, and other innovative products present a real prospect for zero-discharge bioresource utilization.

8.2 Water recovery and reuse in the food and agricultural processing industries

Food and agricultural processing industries use a large amount of fresh water in their operations. As a general observation, the wastewater streams from these operations tend to be non-toxic, even though they are high in COD and TSS, and this is particularly true with wastewater generated from the food industry. Water condensates from water vapor and steam exhausts can be simply filtered, disinfected and reused in the plant. Other sources of wastewater may require many conventional wastewater treatment processes to reach the quality of potable water or other reusable waters, as described in previous chapters.

8.3 CARBOHYDRATES, FATS, AND PROTEINS FOR HUMAN AND ANIMAL

Table 8.3 Potential applications of membrane processes in water recycling and reuse

Sector	Wastewater sources	Potential membrane technology applications
Meat, poultry, and fishery	Slaughterhouse, fish filleting, crustaceans and tuna cooking	MF, UF, NF, RO combined with biological treatment and disinfection
Bakery and grain processing	Grain cleaning, corn wet milling	MF, UF, NF, RO
Fruit and vegetable	Vegetable and fruit rinsing, cleaning, sorting, moving the processed vegetables and fruits, bottle washing, fruit processing	MF, UF, NF, RO
Nut and oilseed	Oil bearing seed milling and washing	MF, UF, NF, RO
Dairy	Vapor condensates, flash coolers, bottle washing and cheese processing	UF, NF, RO with disinfection
Beverage (non-alcohol)	Bottle washing, fruit processing, juice production and cleaning of pipes and tanks	NF, RO
Beverage (alcohol)	Bottle washing, brewing room, biofuel distillation	NF, RO, PV

The wastewaters from certain types of food processing, such as retort and blanching, or cleaning effluents from fruit and vegetable processing/rinsing and bottle washing, may be easily treated and reused. With advances in material science, membrane processes are increasingly used in combination with other unit operations to recycle these types of wastewaters from food and agricultural processing. Table 8.3 lists the potential uses of membrane processes for water recycling of certain types of food and agricultural processing wastewaters.

In many cases, water recycling and reuse of treated water can be combined with the recovery of valuable materials from food and agricultural wastewaters and their final discharges.

8.3 Recoverable carbohydrates, fats, and proteins for human and animal consumption

Currently, whey proteins are recovered from liquid whey for both human consumption and animal feed in limited quantities in the dairy industry. Whey proteins have found uses in infant formulae, in health foods, and in bakery goods. However, some cheese manufacturing plants in which whey is produced as a by-product of cheese making (80% of liquid milk ends up as whey) still do not have the technology (ultrafiltration) to process it. Only 70% of the total available whey is sold as value-added commodity. In addition to unfamiliarity with the technology and the costs associated with implementation of the technology to recover whey proteins, the supply and demand issue also plays a role in this imbalance. The problem can only be solved through the discovery of new uses of whey proteins.

The potential applications of recovered whey proteins (either as a whole or as individual components) as edible oxygen-barrier coatings on foods, as grease barrier coating on paper used for the food service industry or gross coating of confectionery (developed by University of California at Davis), and as pharmaceutical intermediate specialty chemicals in the future, look promising. Another trend of research on carbohydrate and protein recovery from dairy wastewaters which has gathered momentum recently is to fractionate dairy wastewaters into lactose-rich and protein-rich streams, using ultrafiltration and nanofiltration processes (Chollangi & Hossain, 2007; Muro et al., 2010). If this type of fractionation-based approach to recovery of valuable constituents of wastewater of the food industry is adopted and is proven to be economically viable, the impact on the industry and the environment will be significant.

Wastewaters generated in seafood and fish processing plants are enormous, but they contain no toxic substances and, thus, they are readily available for recovery of food materials, some of which are soluble proteins. However, recovery of food materials from seafood and fish processing facilities is usually limited to large-scale operations, as wastes from small processors do not have enough recoverable materials to justify the cost of operating even a small batch recovery unit. Where the recovery operation is applicable, though, it is an excellent way to reduce BOD_5 in wastewater and produce additional products to offset wastewater treatment. It is feasible to recover soluble proteins with a properly selected membrane separation process and then process the proteins into flour after the fish oil is removed, while screens and flotation devices can recover fish debris and greases from wastewater streams for used in animal feeds.

As described in Chapter 3, membrane-based technologies have great potential to concentrate, to fractionate and to purify organic and inorganic materials in aqueous solutions or suspensions, or they can be used as intermediate separation steps to facilitate efficient and economical recovery of value-added materials. In earlier times, membrane separations had been used primarily to recycle the wash water in seafood and fish processing operations (Pavia & Tyagi, 1972); the aim of the processing of wastewater from fish processing was not material recovery. Increasingly, however, more attention has been paid to the economical effect of protein or enzyme recovery from fish processing wastewaters (Pedersen et al., 1987, 1989; Afonso & Bórquez, 2002; Dumay et al., 2008).

8.3 CARBOHYDRATES, FATS, AND PROTEINS FOR HUMAN AND ANIMAL

Wastewaters from vegetable and fruit processing operations represent a large segment of food and agricultural wastewaters that can be utilized. The amount of water used for these operations is huge, and so is the potential impact of its wastewaters on downstream municipal wastewater treatment plants. One example of common vegetable processing is potato processing.

Potato processing operations produce a large amount of wastewater effluent high in COD, TTS, and Total Kjeldahl Nitrogen or TKN, a measurement of protein content. The major source of the effluent from a typical potato processing plants is from:

1. Potato washing: dirt or silt adhering to the surface of the potato is removed in a washing step, and the water from this process contributes to the suspended solids load of the effluent. Most of this material is non-organic in nature.

2. Peeling operation: the potato skin is removed, which generates a lot of suspended solids and 'grey' starchy water. This step contributes to both the organic and suspended solids loads of the effluent.

3. Slicing and trimming of potatoes: during the potato slicing process, starch mixed with trims and sands contributes to the suspended solids and organic load of the effluent. Small pieces of potato slices also end up in the effluent, adding to the organic load.

4. Blanching (optional for potato chips): blanching in chip making reduces the transfer of broken pieces of potato into the fryer, while also reducing fryer energy usage and eliminating unevenly-fried chips. It contributes to the organic load as well as the suspended solids load of the effluent.

5. Boil-out operations: during a boil-out operation, the fryers are sanitized with caustic soda and water, contributing to the suspended solids – the organic content, as well as the fats, oils and grease (FOG) content of the effluent.

All of these processes use large volumes of water, which generate the same amount of effluent. The effluent contains a high content of both starch (from washing, peeling, slicing, and blanching) and oil (from frying). Recovery of these materials can produce benefits of low COD wastewater suitable for directly discharging into a municipal sewage system, and also useful materials that can be sold for profit (e.g., as animal feeds). Catarino et al. (2007) developed an integrated system to recover starch from wastewaters collected before frying, using a series of hydrocyclones and a vacuum filter unit, where grease and oil were removed from a gravity settling tank.

If one chooses to treat the effluent without attempting to recover the organic materials, anaerobic treatment of potato processing wastewater is often

Figure 8.1 Sludge from anaerobic digestion of potato processing wastewater. *Source*: Steven Vaughn, USDA ARS in Peoria, IL, USA

employed (Zoutberg & Eker, 1999). The resulting sludge from an anaerobic digester of potato processing wastewater treatment could be used as soil amendment, as dried sludge has similar characteristics to peat moss (see Figure 8.1), a diminishing natural resource that is widely used in landscaping and horticulture.

8.4 Recoverable aroma flavoring compounds from food processing

Recovery of flavors and savory compounds from food wastewaters has long been contemplated, but it was not economically and technically possible (with few exceptions) until the advent of membrane-based technologies. Today, pervaporation technology has been used successfully for recovering flavor products from fermentation broth in bioreactors and for aroma recovery (Trifunovic & Trägardh, 2002; Peng & Liu, 2003).

Wastewaters from fruit and vegetable processing operations, in particular, are good candidates for flavor recovery. For example, blanching waters from vegetables processing contain aroma compounds that could be harvested for use in a number of food and non-food applications. Blanching is the process of heating vegetables to a temperature high enough to destroy enzymes (and possibly some

Table 8.4 Abundant volatile flavor compounds in commonly blanched vegetables

Broccoli	Brussels sprouts	Cauliflower	Beans
5-(methylthio) pentanenitrile	2-propenyl isothiocyanate	Dimethyl sulfide	1-octen-3-ol
4-(methylthio)butyl isothiocyanate	Dimethyl sulfide	Dimethyl disulfide	(Z)-5-octen-2-one
Dimethyl sulfide	Dimethyl disulfide	(S)-methyl thio-butyrate	(Z)-3-hexenol
Dimethyl disulfide	(Z)-3-pentenol	acetone	n-hexanol
Nonanal	acetaldehyde	acetaldehyde	acetaldehyde
(E,E)-2,4-heptadienal	Diethyl ketone	Allyl cyanide	Dimethyl disulfide
(Z)-3-hexenol	acetone	Trans-but-2-een-1-ol	Ethyl alcohol
Allyl cyanide	Allyl cyanide	Allyl isothiocyanate	Diethyl ketone

Source: Data from MacLeod & MacLeod (1970), Maruyama (1970), Buttery *et al.* (1976), and Whitfield & Last (1991).

microorganisms) present in the tissue, and it stops the enzyme action that causes loss of color and flavor during storage. In water blanching, the vegetables are submerged in boiling water. In steam blanching, the vegetables are suspended above the boiling water and heated only by the steam. Water blanching usually results in a greater loss of nutrients through leaching, but it takes less time than steam blanching. The blanched foods are usually submerged in cooling water immediately after blanching. The combined wastewater from blanching and cooling contain amounts of soluble solids, coarse debris, fine particles, and small molecules such as odorous (aroma) compounds.

The aroma compounds in blanching water can vary depending on the vegetable being blanched. For example, more than 33 aroma compounds have been found in cooked cauliflower, cabbage, Brussels sprouts and runner beans (MacLeod & MacLeod, 1970). Some of the aroma compounds that could be present in the blanching water of several common vegetables is listed in Table 8.4. The flavors listed are either characteristic in cooked (blanched) vegetables or are relatively abundant in terms of approximate percentages. These compounds can all be recovered with membrane pervaporation technology; however, separation or fractionation of these molecules is difficult at present.

8.5 Recoverable food/agricultural biomaterials for non-food uses

As shown in Table 8.2, specialty chemicals such as tartarate can be produced from wine grape wastes, while glue and gelatin can be extracted from wastes from meat and poultry processing operations. Dyes, as well as cosmetics and nutraceuticals, can be obtained from vegetable and fruit processing. Hormones and vitamins may also be produced from meat, poultry and fishery processing.

Using wastewaters from food processing as substrates for various bioproductions of potential economic interest has been reported (Murado et al., 1993; Roukas, 1999; Guerra & Pastrana, 2003). For example, wastes from mussel processing have been used to produce single cell protein and a highly stable amylolytic preparation from different *Aspergillus* strains (Murado et al., 1993). Recently, Hernández et al. (2006) have produced amylase and protease by *Aspergillus niger* strain UO-1, using the wastewaters from brewery and meat processing. Amylase is widely used in the food industry, ethanol brewing in general and in the textile and paper industries, while protease can be used in cheese production, meat tenderizer, baking, textiles, tanning of leathers, and also as an additive to detergents.

In addition to ethanol or methane, which can be converted from recovered starch from wastewaters and used either as a fuel or as a chemical, there are many examples of specialty chemical products that result from fermentation of carbohydrates by different species of microorganisms (Leeper et al., 1991). Lactic acid, succinic acid, aliphatic acids, sorbitol, stearic acid, and linoleic acid are just a few of many possible chemical products from fermentation of sugars (glucose, sucrose, and lactose).

Other than membrane processes and filtration, coagulation/flocculation is commonly used in recovering materials from food and agricultural processing wastewaters. For example, Khwaja & Vasconcellos (2011) developed a method for recovering tallow from meat processing wastewater, including adding a coagulant composition (a tannin-containing polymer) to the wastewater to agglomerate suspended fat, oil and grease particles in the wastewater, separating solid waste materials from the wastewater, and isolating tallow from the solid waste materials. Other coagulants or coagulant aids, such as lignin, alginate, and other natural or synthetic polymers, have been used in removing proteins and fats from food processing wastewaters (Coffman et al., 1988; Miller, 1996). For extraction of bioactive compounds or other small molecules, solvents and supercritical CO_2 technology may be used to extract phytochemicals or useful compounds from filtrates and other concentrated waste materials separated from food and agricultural wastewaters.

8.6 Energy or fuel generations from wastewaters

Ethanol is now being used in gasoline blends and fuel for specifically designed automobile engines. Ethanol can be produced from food and agricultural

8.6 ENERGY OR FUEL GENERATIONS FROM WASTEWATERS

wastewaters as long as there are sufficient amounts of sugar or starch present. The fermentation-produced ethanol has a relatively low ethanol content, which has to be enriched to 95% or higher for used as fuels for internal combustion engines. A combination of distillation and pervaporation will produce almost 100% pure ethanol (Peng *et al.*, 2003).

Biogas from anaerobic processes, such as anaerobic sludge digesters or anaerobic reactors for reducing high-strength wastewaters, has been well known and utilized to some degree on a small scale. However, the enthusiasm for its energy generation capacity has never lasted very long, as people soon realize the costs associated with enriching methane gas from biogas, and the collection and transportation of this gas in such small quantities.

Landfills produce biogas naturally under anaerobic conditions, but this gas has attracted little serious notice until recently, as groups interested in the biogas from landfills have shared little common ground with one another. Those anxious about global warming have been concerned about the fact that methane and CO_2 comprise of the majority of biogas from landfills, while entrepreneurs have seen the same biogas as "diamond in the rough" – part of a new "green revolution" that will usher in a green economy.

In December 2008, Cavendish Farms, a Canadian potato farming and processing company, built a biogas facility in Charlottetown, Prince Edward Island (See Figure 8.2). The facility takes solid waste material from potato processing and its wastewater and, through anaerobic digestion (using proprietary microorganisms), converts it into energy for the potato processing plant. This is the first facility in the potato industry to utilize solid potato waste and convert it into usable energy for the operation of the facility itself. The digestate from this conversion facility has been demonstrated to be useful as horticultural substrate (peat moss substitute). Figure 8.3 shows the growth difference among tomato plants planted in the potato digestate, in regular compost plus vermiculite, and in peat moss plus vermiculite. The potato digestate outperformed the other substrates.

Figure 8.2 Cavendish biogas facility. *Source*: Steven Vaughn, USDA ARS in Peoria, IL, USA

Figure 8.3 Red Robin tomato plants grew in different substrates after four weeks. *Source*: Steven Vaughn, USDA ARS in Peoria, IL, USA

Biogas is not the only combustible gas produced from treatment processes of food and agricultural wastewaters. Research on producing hydrogen gas through fermentation of sugars in wastewaters has shown promising results (Ueno *et al.*, 2007; Van Ginkel *et al.*, 2005; Hussy *et al.*, 2005). Although hydrogen production from fermentation of wastewaters reduces no significant amount of solid content, another approach, which combines hydrogen gas production and electricity generation in a single system, is gaining new attention and interests because of its waste reduction as well as energy generation (Logan, 2004; Oh & Logan, 2005). The amount of electricity generated in the microbial cell was small, but the potential of the application of this technology looks bright.

Pyrolysis, particularly fast pyrolysis, offers the opportunity to obtain heating oil from solid or semi-solid biomass from food and agricultural wastewaters. Pyrolysis is a thermal decomposition of organic compounds in the absence of oxygen. This process has been used for hundreds of years to produce charcoal, and has been applied commercially to recover methanol, acetic acid, and turpentine. The process of fast pyrolysis produces a liquid whose yield depends on the biomass composition and the rate and duration of heating and a char. Although the liquid (which contains up to 15–20% water) looks like oil and is called "bio-oil," the elemental composition of "bio-oil" resembles that of the biomass rather than that of typical petroleum oils, with highly oxygenated compounds providing lower heat values. Because of several organic acids in the pyrolysis liquid, the liquid is very corrosive. The char can be used as adsorbent and the components in the liquid can, theoretically, be separated in order to achieve their full potential as specialty chemicals. However, the problem of chemical or physical separations is formidable in both the technical and economical senses.

If the biomass is sludge, the pulverized sludge is heated to 250°C and compressed to 40 MPa. The hydrogen in the water inserts itself between chemical bonds in natural polymers such as fats, proteins and cellulose. The oxygen in the water combines with carbon, hydrogen and metals. The result is oil, light combustible gases such as methane, propane and butane, water with soluble salts, carbon dioxide, and a small residue of inert insoluble material that resembles powdered rock and char. This process, which uses hydrous pyrolysis to convert reduced complex organics to oil, is called thermal depolymerization.

All organisms and many organic toxins are destroyed. Inorganic salts such as nitrates and phosphates remain in the water after treatment at sufficiently high levels to require further treatment. The energy from decompressing the material is recovered, and the process heat and pressure is usually powered from the light combustible gases. The oil is usually treated further to make a refined useful light grade of oil, such as No. 2 diesel and No. 4 heating oil.

8.7 Algae-based biodiesel and fuel ethanol

Algae growth in wastewater treatment plants, with a few exceptions such as certain applications of stabilization ponds, are generally undesirable, as algae clogs the treatment system, is difficult to settle, interferes with disinfection, increases biomass, and competes with other microorganisms in biological treatment processes. Also, it is aesthetically challenging for many people.

Recently, interest in algae based biodiesel and biofuel production has led to the development of new technologies to harness the rich nutrients in food and agricultural wastewaters to produce oil- and or/carbohydrate-rich algae for alternative energy productions. As a biomass, algae can be utilized in a number of ways to produce biofuel, such as gasification, pyrolysis, hydrogenation, fermentation, anaerobic digestion and liquefaction of algal biomass to produce gas- or oil- based biofuel (McKendry, 2002a, 2002b; Miao & Wu, 2006; Amin, 2009; Woertz *et al.*, 2009; Brennan & Owende, 2010; Rawat *et al.*, 2011; Pittmann, 2011; Park *et al.*, 2011; Gendy & El-Temtamy, 2013).

Green unicellular microalgae, such as *gracilaria, botryococcus, dunliella teriolecta, sargassum, and chlorella*, are the most promising algae for renewable energy production. It is said that microalgae have a higher biomass production rate than other biofuel crops in terms of land area used and costs. Many new designs of microalgal systems have the growth substrate gone vertical, which further reduces the land area required for microalgae production. The production of microalgae in agricultural and food wastewater requires no fresh water input, thus eliminating one of the common limiting factors for the cultivation of many plant-based biofuel crops.

The efficiency of algal growth in wastewater is highly dependent on temperature, pH, and growth media in the wastewater stream. The growth media for growth of microalgae includes nutrients (e.g., nitrogen and phosphorus (N and P)), but will also contain other microorganisms that might compete with the microalgae (or/and in the case of zooplanktons, prey on microalgae) used for biofuel productions. There have been many studies that focused on the growth media (e.g., Shi et al., 2007; Bhatnagar et al., 2010).

As wastewater from one site could have different growth media from another site, there will be variations in species of microalgae used for microalgae production. For wastewater steams rich in nitrogen and phosphorus found in many agricultural and food wastewaters, the issue of N- and P-tolerant microalgae species becomes very important. It has been widely reported that microalgae species such as *Chlorella* and *Scenedesmus* genus are particularly tolerant to the high concentrations of nitrogen and phosphorus that are often found in many tertiary (advanced) treatment processes in agricultural, food, and municipal wastewater treatment plants, and can effectively remove large percentages of N and P from wastewater streams.

There are three types of biofuels that can be produced from conversion of microalgae growing in agricultural and food wastewaters – bioethanol, biodiesel, and biomethane:

- Bioethanol is a modified version of ethanol production from biomass. Either biochemical conversion (fermentation) or a thermo-chemical process such as gasification is used to produce bioethanol. The fermentation process generates ethanol, CO_2, and methane (methane is produced from digestion of the remaining algae after fermentation) (Gendy et al., 2013).

- Biodiesel is obtained from lipids in microalgae that undergo transesterification, which is similar to biodiesel production from plant-based oils.

- Biomethane is biogas (methane along with CO_2) produced in anaerobic digestion of microalgae or related biomass, and it can be used for fuel gas for heating and generating electricity. Comparing to other biomass, microalgae lack lignin and have less cellulose, making them efficient feedstock for conversion.

The current research on microalgae-based biofuel from wastewaters is still more or less a bench scale affair, and many challenges are yet to be overcome. However, a niche opportunity may exist for a particular type stabilization pond treatment of wastewater called high rate algal ponds (HRAPs) (Park et al., 2011). These are aerobic ponds that were developed in 1950s as an alternative to unmixed oxidation ponds for BOD_5, suspended solids, and pathogen removal

8.8 POTENTIAL APPLICATIONS OF INDUSTRIAL COMMODITIES DERIVED   239

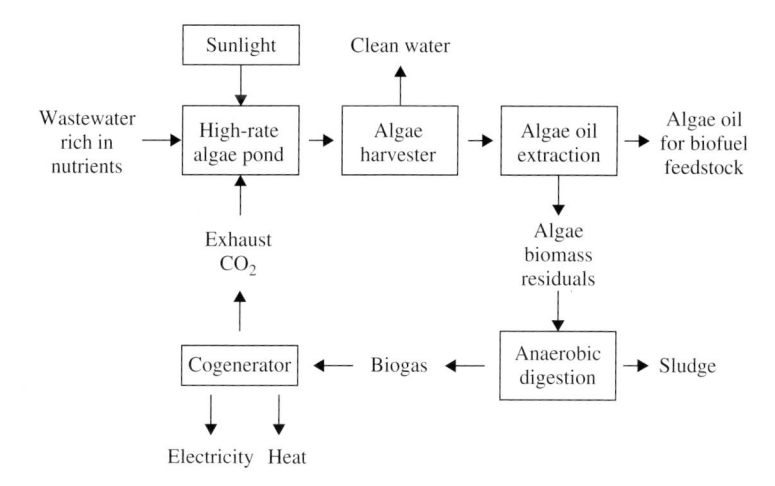

Figure 8.4 Flow diagram envisioned for algae wastewater treatment and liquid biofuel production. *Source*: Woertz *et al.* (2009). With permission of ASCE

(Rawat *et al.*, 2011). HRAPs are shallow, open raceway ponds of wastewater treatment facilities, where microalgae are produced, harvested, and transformed into bioethanol, biomethane, and biodiesel, and this is presently the only commercially viable system of algal biomass production and subsequent conversion to bioenergy with minimum footprint and environmental impact (Park *et al.*, 2011).

Woertz *et al.* (2009) in Cal Poly proposed an algae wastewater treatment and liquid biofuel production process (see Figure 8.4). They provided a proof-of-concept for HRAP process that combined N and P removal with algal lipid production for both dairy and municipal wastewaters. Carbon dioxide was used to accelerate the growth of microalgae in both outdoor and indoor mixed species cultures (with the additional benefit of CO_2 sequestration). These authors found that lipid contents ranged from 4.9% to 29%, and lipid productivity peaked at 2.8g/m²/d. In addition to CO_2, other strategies need to be applied to optimize microalgae production. Species control and control of grazers and parasites are the most important practical options to enhance algal production and/or lipid yield (Park *et al.*, 2011).

8.8 Potential applications of industrial commodities derived from sludge treatment

Treated sludges are used beneficially in land application for agriculture as fertilizers or soil conditioners. These practices have been employed for centuries, and other unconventional uses of sludge products have also been vigorously explored around the world. This phenomenon is more prominent outside the United States of America where land available for sludge disposal is limited and ocean disposal of sludges is banned, as in the European Union nations.

Many proposals of potential uses of sludge products have been put forth, and some of them have been tried out in laboratory or in commercial-scale operations. For example, dewatered treated sludges have been used successfully for manufacturing building materials, such as concrete and bituminous mixes, and also as a road subsoil additive, utilizing chemical fixation processes (Aziz & Koe, 1990). The chemical fixation process involves combining treated sludge with stabilizing agents such as cement, sodium silicate, pozzolan (fine-grained silicate), or lime, to chemically react with or encapsulate sludge particles (Metcalf & Eddy, 1991). The process produces a product with a high pH that inhibits viruses and bacteria and, for many chemical treatment products, the product has the consistency of natural clay.

Final residuals of incineration or other thermal processes have also been used to generate road sub-base material or concrete aggregate (Takeda *et al.*, 1989). Since the main ingredients for Portland cement powder are limestone, clay, silica, iron, etc., and incinerated sludge ash usually contains same ingredients as clay, it has been successfully used as a part of the cement materials. Pulverized sludge ash, limestone, and dewatered sludge/clay slurries have been used successfully in lightweight concrete applications without influencing the product's bulk properties (Tay & Show, 1991, 1993). Sludge-based concrete has been deemed suitable for load-bearing walls, pavements, and sewers (Lisk, 1989). The cement industry is highly energy intensive, but the large energy costs of creating clinker (powdered cement produced by heating a properly proportioned mixture of finely ground raw materials in a kiln) at 1500°C can be offset by utilizing sludges as a low-cost and readily available supplemental energy source, depending on the percentage of volatile solids in the sludges. Furthermore, sludges can be injected into the exhaust gas chamber to eliminate NO_x emissions, using the heat of the hot exhaust gases reacted with ammonia contained in the sludge to convert NO_x to nitrogen gas (Kahn & Hill 1998).

Solidification of hazardous materials and heavy metals has long been an effective method to prevent harmful materials from leaching into the environment when these materials are disposed of. This process can also be used for the production of sludge-based products. The sludge from a food processing plant is unlikely to have the metals or other inorganic substances that are needed to produce these materials but, since some streams of food and agricultural wastewater have been discharged into a municipal sewage system (maybe after some forms of treatment to reduce BOD_5 and TSS), the processes described below have some relevance.

It is possible to incorporate sludge enriched with heavy metals into the manufacture of biobricks. In this approach, incinerator sludge ash is used as a clay substitute during the manufacture of these bricks. The process is said to improve the ceramic properties and product strength of the resulting construction materials (Anderson *et al.*, 1996), and it is reassuring to know that these biobricks do not release heavy metals during firing in the production or by weathering when in use (Alleman *et al.*, 1990).

Additional benefits of the biobrick technologies also include volume reduction and substantial savings on water and fuel consumption, as well as treatment costs. In Tokyo, Japan, a product called Metro-Brick, made of 100% incinerated ash through mechanical compression under high temperature (1,050°C), has been used as pavement materials for sidewalks, community roads, public open spaces, and parks. Sludges have also been tried out for use as an "activated carbon" for odorous gas treatment via adsorption and for flue gas treatment via desulfurization (Krogmann *et al.*, 1997). Palantzas & Wise (1994) investigated the possibility of producing calcium magnesium acetate using residual biomass from sewage sludge. It is reported that this technique would generate an overall cost savings of 68% over conventional disposal costs.

A technology called "sludge-to-fuel" (STF) utilizes the volatile solids in many biosolids for producing combustible oil. STF involves a process that converts sludge organic matter into combustible oil, using a solvent under atmospheric pressure, with temperatures ranging from 200–300°C (Millot *et al.*, 1989). Alternatively, STF using hydrous pyrolysis can produce combustible oil under high pressures in the range of 10 MPa and high temperatures (Itoh *et al.*, 1994). One STF system employs a hydrothermal reactor to convert mechanically dewatered sludge to oil, char, CO_2, and wastewater. The char, making up 10% of the product, is sent to a landfill, and the wastewater returns to the wastewater treatment system, while the gaseous emissions are treated and released into the atmosphere. The produced oil has approximately 90% of the heating value of diesel fuel and can be sold to offsite users or refineries (Hun, 1998). This is an example of thermal depolymerization to transform reduced complex organics to oil.

Other STF processes produce oils from sludge by employing activated alumina pyrolysis of digested, dried sludges, or toluene-extracted sludge lipids (Abu-Orf & Jamrah, 1995). In the case of the sludge sources, sludge-associated metals seem to bind to the residuals, with final product conversion efficiency being dependent on the sludge particulate size, temperature, and heating rate (Takeda *et al.*, 1989). Metals in sludges tend to become trapped in the residue of the STF process, while organo-chlorine compounds that survive treatment within a typical sewage treatment plant are likely to be destroyed in the processes (Bridle *et al.*, 1990). Oils produced with the STF technology have the potential to be used as heating oil and possible chemical feedstocks (Boocock *et al.*, 1992). Much lipid-rich sludge from mechanical food wastewater treatment plants appears to be a good candidate for STF technology.

8.9 Further reading

Metcalf & Eddy, Inc. (Tchobanoglous, G. & Burton, F.L). (1991). *Wastewater Engineering, Treatment, Disposal, and Reuse.* 3rd edition. McGraw-Hill, New York, NY.

Brown, R.C. (2003). *Biorenewable Resources: Engineering New Products from Agriculture.* Blackwell Publishing Professional, Ames, IA, USA.

Waldron, K. W. (ed.) (2007). *Handbook of Waste Management and Co-Product Recovery in Food Processing (Volume 1).* Woodhead Publishing, Cambridge, UK.

8.10 References

Abu-Orf, M.M. & Jamrah, A.I. (1995). Biosolids and Sludge Management. *Water Environment Research* **67**, 481–487.

Alfonso, M.D. & Borquéz, R. (2002). Review of the treatment of seafood processing wastewaters and recovery of proteins therein by membrane separation processes – prospects of the untrafiltration of wastewaters from the fish meal industry. *Desalination* **142**, 29–45.

Alleman, J.E., Bryan, E.H., Stumm, T.A., Marlow, W.W. & Hocevar, R.C. (1990). Sludge-amended brick production: applicability for metal-laden residues. *Water Science and Technology* **22**, 309–317.

Amin, S. (2009). Review on biofuel oil and gas production processes from microalgae. *Energy Conversions and Management* **50**, 1834–1840.

Anderson M., Skerratt, R.G. Thomas, J.P. & Clay, S.D. (1996). Case study involving fluidised bed incinerator sludge ash as a partial clay substitute in brick manufacture. *Water Science and Technology* **34**, 507– 515.

Aziz, M.A. & Koe, L.C.C. (1990). Potential utilization of sewage sludge. *Water Science and Technology* **22**, 277–285.

Bhatnagar, A., Bhatnagar, M, Chinnasamy, S. & Das, K. (2010). Chlorella Minutissima – a promising fuel algae for cultivation in municipal wastewaters. *Applied Biochemistry and Biotechnology* **161**, 523–536.

Boocock, D.G.B., Konar, S.K., MacKay, A., Cheung, P.T.C. & Liu, J. (1992). Fuels and chemicals from sewage sludge. 2. The production of alkanes and alkenes by the pyrolysis of triglycerides over activated alumina. *Fuel* **71**, 1291–1297.

Brennan, L. & Owende, P. (2010). Biofuels from microalgae – a review of technologies for production, processing, and extractions of biofuels and co-products. *Renewable & Sustainable Energy Reviews* **14**, 557–577.

Bridle, T.R., Hammerton, I. & Hertle, C.K. (1990). Control of heavy metals and organochlorines using the oil from sludge process. *Water Science and Technology* **22**, 249–258.

Buttery R.G., Gaudagni D.G., Ling L.C., Seifert R.M. & Lipton W. (1976). Additional volatile components of cabbage, broccoli, and cauliflower. *Journal of Agricultural and Food Chemistry* **24**, 829–832.

8.10 REFERENCES

Catarino, J., Mendonça, E., Picado, A., Anselmo, A., da Costa, J.N., & Partidário, P. (2007). Getting value from wastewater: by-products recovery in a potato chips industry. *Journal of Cleaner Production* **15**, 927–931.

Chollangi, A. & Hossain, M. (2007). Separation of proteins and lactose from dairy wastewater. *Chemical Engineering and Processing* **46**, 398–404.

Coffman, G.L., Markham, W.M., & Reid, J.H. (1988). Recovery of fats and proteins from food processing wastewaters with alginates. US Patent 4933087A.

Dumay, J., Radier, S., Barnathan, G., Bergé, J. & Jaouen, P. (2008). Recovery of valuable soluble compounds from washing waters generated during small fatty pelagic surimi processing by membrane processes. *Environmental Technology* **29**, 451–461.

Gendy, T.S. & El-Temtamy, S.A. (2013). Commercialization potential aspects of microalgae for biofuel production: an overview. *Egyptian Journal of Petroleum*, in press.

Guerra, N.P. & Pastrana, L. (2003). Enhancement of nisin production by *Lactococcus lactis* in periodically re-alkalized cultures. *Biotechnology and Applied Biochemistry* **38**, 157–167.

Hernández, M.S., Rodríguez, M.R., Guerra, N.P., & Rosés, R.P. (2006). Amylase production by *Aspergillus niger* in submerged cultivation on two wastes from food industries. *Journal of Food Engineering* **73**, 93–100.

Hun, T. (1998). Compact solids-to-oil process cuts costs. *Water Environment and Technology* **10**, 22–24.

Hussy, I. Hawkes, F. R., Dinsdale, R., Hawkes, D.L. (2005). Continuous fermentative hydrogen production from sucrose and sugarbeet. *International Journal of Hydrogen Energy* **30**, 471–483.

Itoh, S., Suzuki, A., Nakamura, T. & Yokoyama, S. (1994). Production of Heavy oil from sewage sludge by direct thermochemical liquefaction. *Desalination* **98**, 127–133.

Kahn, R. & Hill, P. (1998). An uncommon use: county sanitation districts of Los Angeles county sends biosolids to a cement plant to reduce manufacturing emissions. *Water Environment and Technology* **10**, 44–49.

Khwaja, A.R. & Vasconcellos, S.R. (2011). Methods for recovering tallow from wastewater. US Patent 7943048 B2.

Krogmann, U., Boyles, L.S., Martel, C.J. & McComas, K.A. (1997). Biosolids and sludge management. *Water Environmental Research* **69**, 534–550.

Leeper, S.A., Ward, T.E., & Andrews, G.E. (1991). *Production of Organic Chemicals via Bioconversion: A Review of the Potential*. Report EGG-BG-9033, Idaho National Engineering Laboratory, Idaho Falls, ID, USA.

Lisk, D.J. (1989). Compressive strength of cement containing ash from municipal refuse or sewage sludge incinerators. *Bulletin of Environmental Contamination and Toxicology* **42**, 540–543.

Logan, B.E. (2004). Biologically extracting energy from wastewater: biohydrogen production and microbial fuel cells. *Environmental Science and Technology* **39**, 5488–5493.

MacLeod A.J. & MacLeod, G. (1970). Flavor volatile of some cooked vegetables. *Journal of Food Science* **35**, 734–738.

Maruyama, F.T. (1970). Identification of dimethyl trisulfide as a major component of cooked brassicaceous vegetables. *Journal of Food Science* **35**, 540–543.

McKendry, P. (2002a). Energy production from biomass (part 2): conversion technologies. *Bioresource Technology* **83**, 47–54.

McKendry, P. (2002b). Energy production from biomass (part 3): gasification technologies. *Bioresource Technology* **83**, 55–63.

Metcalf & Eddy, Inc. (Tchobanoglous, G. & Burton, F.L). (1991). *Wastewater Engineering, Treatment, Disposal, and Reuse*. 3rd edition. McGraw-Hill, New York, NY.

Miao, X.L. & Wu, Q.Y. (2006). Biodiesel production from heterotrophic microalgal oil. *Bioresource Technology* **97**, 841–846.

Miller, J.C. (1996). Process for removing proteinaceous materials, fat and oils from food processing wastewater and recovering same. US Patent 5543058.

Millot, N., Huyard, A., Faup, G.M. & Michel, J.P. (1989). Sludge liquefaction by conversion to fuels. *Water Science and Technology* **21**, 917–923.

Murado, M.A., Siso, M.I.G., González, M.P., Montemayor, M.I., Pastrana, L. & Miron, J. (1992). Characterization of microbial biomasses and amylolytic preparations obtained from mussel-processing waste treatment. *Bioresource Technology* **43**, 117–125.

Muro, C., Díaz, C., García, B., Zavala, R., Ortega, R., Álvarez, R, & Riera, R. (2010). Recuperación de Los Components Del Lactosuero Residual de Una Industria Elaboradora de Queso Utlizando Membranas. *Afinidad: Revista de Química Teórica Y Aplicada* **67**, 212–220.

Magbunua, B. (2000). *An Assessment of the recovery and Potential of Residuals and By-Products from the Food Processing and Institutional Food Sectors in Georgia*. University of Georgia Engineering Outreach Services, Athens, GA.

Oh, S. & Logan, B.E. (2005). Hydrogen and electricity production from a food processing wastewater using fermentation and microbial fuel cell technologies. *Water Research* **39**, 4673–4682.

Palantzas, I.A. & Wise, D.L. (1994). Preliminary economic analysis for production of calcium magnesium acetate from organic residues. *Resource Conservation and Recycling* **11**, 225–231.

Pavia, E.H. & Tyagi, A.D. (1972). *Safe Economical Rescue of Transport Water in the Fish Meal and Oil Industry*. ASME, New York, NY.

Park, J.B.K., Craggs, R.J., & Shilton, A.N. (2011). Wastewater treatment high rate algal ponds for biofuel production. *Bioresource Technology* **102**, 35–42.

Peng, M., Vane, L.M., & Liu, S.X. (2003). Recent advances in voc removal from water by pervaporation. *Journal of Hazardous Materials* **B98**, 69–90.

Peng, M. & Liu, S.X. (2003). Recovery of aroma compound from dilute model blueberry solution by pervaporation, *Journal of Food Science* **68**(9), 2706–2710.

Pedersen, L.D., Rose, W.W., Brewbaker, D.L. & Merlo, C.A. (1987). *Hyperfiltration Technology for Recovery and Utilization of Protein Materials in Surimi Process*

Waters, Final Report Phase I. National Food Processors Association, Dublin, CA, USA.

Pittman, J.K., Dean, A.P. & Osundeko, O. (2011). The potential of sustainable algal biofuel production using wastewater resources. *Bioresource Technology* **102**, 17–25.

Pedersen, L.D., Rose, W.W., Deniston, M.F. & Merlo, C.A. (1989). *Hyperfiltration Technology for Recovery and Utilization of Protein Materials in Surimi Process Waters, Final Report Phase II.* National Food Processors Association, Dublin, CA, USA.

Rawat, I., Kumar, R.R., Mutanda, T., & Bux, F. (2011). Dual role of microalgae: phycoremediation of domestic wastewater and biomass production for sustainable biofuels production. *Applied Energy* **88**, 3411–3424.

Roukas, T. (1999). Pullulan Production from brewery wastes by *Aureobasidium polluans*. *World Journal and Biotechnology* **106**, 23–32.

Shi, J., Podola, B. & Melkonian, M. (2007). Removal of nitrogen and phosphorus from wastewater using microalgae immobilized on twin layers: an experimental study. *Journal of Applied Phycology* **19**, 417–423.

Takeda, N., Hiraoka, M., Sakai, S., Kitai, K. & Tsunemi, T. (1989). Sewage sludge melting process by coke-bed furnace system development and application. *Water Science and Technology* **21**, 925–935.

Tay, J.H. & Show, K.S. (1991). Properties of cement made from sludge. *Journal of Environmental Engineering* **117**, 236–246.

Tay, J.H. & Show, K.S. (1993). Manufacture of cement from sewage sludge. *Journal of Materials in Civil Engineering* **5**, 19–29.

Trifunovic O. & Trägardh G. (2002). Transport of diluted volatile organic compounds through pervaporation membranes. *Desalination* **149**, 1–2.

Ueno, Y., Tatara, M., Fukui, H., Makiuchi, T., Goto, M. & Sode, K. (2007). Production of hydrogen and methane from organic solid wastes by phase-separation of anaerobic process. *Bioresource Technology* **98**, 1861–1865.

Van Ginkel, S. W., Oh, S. & Logan, B.E. (2005). Biohydrogen gas production from food processing and domestic wastewaters. *International Journal of Hydrogen Energy* **30**, 1535–1542.

Woertz, I., Feffer, A., Lundquist, T. & Nelson, Y. (2009). Algae grown on dairy and municipal wastewater for simultaneous nutrient removal and lipid production for biofuel feedstock. *Journal of Environmental Engineering (ASCE)* **135**, 1115–1122.

Whitfield, F.B. & Last, J.H. (1991). Vegetables. In: Maarse, H. & Dekker, M. (eds.) *Volatile Compounds in Foods and Beverages.* pp. 203–281. New York, NY.

Zoutberg, G.R. & Eker, Z. (1999). Anaerobic treatment of potato processing wastewater. *Water Science and Technology* **40**, 297–304.

9

Economics of food and agricultural wastewater treatment and utilization

9.1 Introduction

In designing a wastewater treatment system for an onsite wastewater treatment plant, the primary criterion for selection of one design over another is protection of public health and the health of the workers, while preventing environmental degradation. Secondary criteria are the cost and ease of operating and maintaining the system. The fate of any residuals resulting from the treatment and disposal system must be considered in the selection process. Although there are various wastewater treatment options available for any specific wastewater problem, it is rather difficult to make a decision on the selection of a design option that prevents public health hazards while also maintaining environmental quality at the least cost.

The first step in the design of an onsite system is the selection of the most appropriate components to make up the system. Since the site characteristics constrain the method of disposal more than other components, the disposal component of a wastewater treatment system must be selected first.

To select the disposal method properly, a detailed site evaluation is required. However, the site characteristics that must be evaluated during the design stage may vary with the disposal method. Since it is neither economical nor practical to evaluate a site for every conceivable system design, the purpose of this first step is to eliminate those disposal options that have the least potential, so that the detailed site evaluation can concentrate on the most promising options. To screen the disposal options effectively, the wastewater stream to be treated and disposed must be characterized in as much detail as possible and an initial site investigation made.

The estimated daily wastewater volume and any short- or long-term variations in flow affect the size of many of the system components. In addition, the concentrations of various constituents can affect the treatment and disposal options to be chosen.

For a process designer or an engineer, which process component or processes to apply, and the design of each process based on the characteristics of the wastewater and process calculations and onsite evaluation, are the preliminary factors to consider. After these, economic considerations are among the most important parameters that affect the final decision as to which process or processes should be chosen for wastewater treatment.

In order to estimate the costs of the processes in consideration, the data from the wastewater characterization should be available, along with the design parameters for the processes and the empirical cost correlations for these processes. Costs related to alternative processes and information on the quality of effluent should also be collected prior to the development of cost estimation, in compliance with the regulations regarding wastewater discharge.

9.2 Estimating the unit cost of treating food and agricultural wastewater

Cost estimation of a yet-to-be-built wastewater treatment facility is difficult. The responsibility of an engineer or process designer is not to try to outsmart the experts that are hired eventually to do cost forecasting when the decision to build a specific design of wastewater treatment facility is finalized but, rather, to estimate the total costs of a particular wastewater project in order to compare it to other treatment/management alternatives or options. The estimation of costs in this exercise could be as much as 30% off from the actual costs, which is not unusual. The total costs of building and operating a wastewater treatment facility consist of capital costs and operating costs.

9.2.1 Capital costs

These include the unit construction costs, the land costs, the cost of the treatment units, and the cost of engineering, administration and contingencies. The location should be carefully evaluated in each case, because it affects the capital costs more than the operating costs. The cost of equipment may also be a significant portion of the capital costs in more automated and elaborate installations. US EPA (1983) compiled a list of construction costs for most common unitary processes of wastewater treatment, and some of the cost correlations are

9.2 ESTIMATING THE UNIT COST OF TREATING FOOD AND AGRICULTURAL

Table 9.1 Construction costs for selected unit operations of wastewater treatment

Liquid stream	Correlation
Preliminary treatment	$C = 5.79 \times 10^4 \times Q^{1.17}$
Flow equalization	$C = 1.09 \times 10^5 \times Q^{0.49}$
Primary sedimentation	$C = 1.09 \times 10^5 \times Q^{1.04}$
Activated sludge	$C = 2.27 \times 10^5 \times Q^{0.17}$
Rotating biological contactor	$C = 3.19 \times 10^5 \times Q^{0.92}$
Chemical addition	$C = 2.36 \times 10^4 \times Q^{1.68}$
Stabilization pond	$C = 9.05 \times 10^5 \times Q^{1.27}$
Aerated lagoon	$C = 3.35 \times 10^5 \times Q^{1.13}$
Chlorination	$C = 5.27 \times 10^4 \times Q^{0.97}$

Solids stream	Correlation
Sludge handling	$C = 4.26 \times 10^4 \times Q^{1.36}$
Aerobic digestion	$C = 1.47 \times 10^5 \times Q^{1.14}$
Anaerobic digestion	$C = 1.12 \times 10^5 \times Q^{1.12}$
Incineration	$C = 8.77 \times 10^4 \times Q^{1.33}$

Q = flow rate of raw wastewater stream (millions of gallons per day).
Source: U.S. Environmental Protection Agency 1983.

shown in Table 9.1. These unitary cost estimates were developed for municipal wastewater treatment and may not be totally suitable for food and agricultural wastewater treatment plants or small scale wastewater treatment plants. However, they are quite useful for preliminary estimation and process comparison among different alternatives.

Wright & Woods (1993) compiled capital costs for several processes of physical treatment, where capital cost correlations are given for oil–water separators, equalization basins, primary clarifiers, secondary clarifiers, reverse osmosis and ultrafiltration units, gravity filters, and microscreens. Data are included for raw sewage, intermediate and recirculation pumping stations, for preliminary treatment (including bar screens, grit removal, overflow and bypass chamber and Parshall flume), and for grit removal, comminution and gas stripping. They also compiled capital cost estimations for several biological wastewater treatment processes, where figures are given for aeration basins, mechanical aerators, diffused aeration, conventional activated sludge process, extended aeration, contact stabilization, oxidation ditch, rotating biological contactor, trickling filter, aerobic lagoons, facultative lagoon, aerated lagoons and liners (Wright & Woods, 1994). Similar capital cost estimations for several chemical processes in wastewater treatment were also provided in a later paper (Wright & Woods, 1995). Other useful sources on estimation of capital costs can be found in other literature (Peters & Timmerhaus, 1980; Ulrich, 1984; Brown, 2003). However, some of these cost estimations are based on the process chemical industry and may not be entirely relevant to wastewater treatment plant design and economics.

9.2.2 Operating costs

Operating costs are annually based costs that are required to operate the constructed facility, including both direct and indirect costs. If a loan is taken out in order to construct the wastewater treatment facility, the operating costs should also include capital charge or financial charge. Direct costs consist of chemicals, supplies, and materials, labor, utilities (mainly energy), maintenance, and repairs. Indirect costs comprise overhead, local taxes, and insurance. Chemicals, supplies, and materials are those used in chemical and/or advanced treatment processes, depending on the implementation of the treatment processes and the throughput of the each process.

Labor costs represents wages for personnel who operate the facility. The estimate of the cost can be estimated on the basis of two operators per unit per shift (Ulrich, 1984). However, if the operation in the wastewater treatment facility is more or less labor-intensive, the cost estimates need to change accordingly. Additionally, supervisory labors and clerks should be counted separately as 10–20% of the labor cost.

Utilities include electricity, natural gas or heating oil, potable water, and steam. Electricity by far accounts for the majority of the costs for utilities. Maintenance and repairs typically represent 2–10 % of fixed capital, depending on how reliable and how complex the equipment in each unit is. Fixed capital is the investment in the construction of the wastewater treatment facility, which cannot be recovered easily once spent. If a patent is involved in the process, a royalty has to be paid yearly and the cost of this royalty should be included as a part of direct costs.

Overhead, local taxes, insurance, and general expenses make up the indirect costs. Overhead costs represent fringe benefits (mainly medical, dental, and life insurance), Social Security and Medicare taxes (USA only: other nations may have different but similar social benefit programs), and retirement obligations. Depending on the locality and composition of employees, the cost of overhead could be as high as 70% of costs for labor, supervisory and clerical labor, and maintenance and repairs. Local taxes are difficult to pinpoint, and should be decided on case-by-case basis; 1–2% of fixed capital is suggested as a rough, initial estimate. In the USA, insurance cost can account for 0.4–1% of fixed capital. General expenses include administrative and other corporate expenses, and these can be estimated as 15% of labor cost.

Table 9.2 A summary of operating costs in wastewater treatment facilities

Direct costs	Estimation
Chemicals, supplies, and materials (CSM)	$C_R(\$/kg) \times m(kg/s) \times 31.5 \times 10^6 s/yr \times f_0(hr/yr)$
	Where: C_R is unit cost of CSM; m is feed rate; f is capacity factor
Labor cost	10–15% of fixed capital
Supervisory and clerical labor	10–12% of labor cost
Utilities	Dependent on current energy market
Maintenance and repairs	2–10% of fixed capital
Patent royalty payment	Up to 3% of other direct expenses
Direct subtotal	Sum of all direct expenses
Indirect costs	Estimation
Overhead	50–70% of labor, supervisory/clerical labor, and maintenance and repairs
Local taxes	1–2% of fixed capital – may vary greatly
Insurance	0.4–1.0% of fixed capital
General expenses	15% of labor + 5% of direct expenses
Capital charge	Annual payment of interest and principle on loan: $C_{payment} = C_{loan} i(1+i)^n/[(1+i)^n - 1]$, where n is duration of loan in years
Annual operating costs	Sum of direct costs and indirect costs

Source: Brown (2003). Reproduced with permission of John Wiley & Sons.

A loan of capital may be required to construct a wastewater treatment facility. Depending on the interest rate and the number of the years of the loan period, the annual capital charge could run as high as 20% of the total capital and may impact significantly on operating costs.

The main factors that influence the costs of operation and maintenance are: energy costs; labor costs, including the personnel for operation; maintenance and administrative services; material and chemical costs; capital charge; and the cost of transportation of sludges for final disposal and discharge of treated wastewater. The relative importance of these cost items vary significantly with the location, the quality of the effluent discharged, and the specific characteristics of the wastewater. A summary of operating costs is provided in Table 9.2.

9.2.3 Estimation of total costs

Total costs of a wastewater treatment plant are the totals of capital costs and operating costs. In case capital and operating costs are difficult to estimate, a shortcut formula for small wastewater treatment plants can be employed to save the time to estimate total costs. If one knows the capital costs of a similar plant with a different capacity, the capital costs of the plant of interest can be estimated through cost scaling. The idea originates from the chemical industry and recognizes that capital costs are strong related to equipment size. It further follows that capital costs are proportional to 2/3 power of the ratio of their capacity. This fundamental idea can be expressed as (Ulrich, 1984):

$$C_A = C_B \text{ (Capacity of plant A/Capacity of plant B)}^n \tag{9.1}$$

where:
    C_A = predicted capital costs of plant A
    C_B = known capital costs of baseline plant B
    n = economy of scale sizing component (<1). For capital cost estimation, $n = 2/3$.

Operating costs can also be estimated by an expression similar to Equation (9.1):

$$O_A = O_B \text{ (Capacity of plant A/Capacity of plant B)}^n \tag{9.2}$$

where:
    O_A = predicted operating costs of plant A
    O_B = known operating costs of baseline plant B
    n = economy of scale sizing component (<1). For operating cost estimation, $n = 0.85$.

An alternative procedure for the development of cost models for wastewater treatment systems includes the preparation of kinetic models for the possible treatment alternatives, in terms of area and flow rates at various treatment efficiencies, followed by the computation of mechanical and electrical equipment, as well as the operation and maintenance costs as a function of the flow rates (Uluatam, 1991). Models so developed can be used to select the most appropriate treatment process.

    For a more completely user-friendly and computerized cost estimation for wastewater treatment plants, there are several companies marketing commercial software tools for designers of wastewater treatment plants and those who contemplate installing or running a wastewater treatment facilities onsite. One of these companies is Hydromantis, which recently released *CapdetWorks*® version 2.1.

    CapdetWorks® is a planning level design and costing tool that allows the user to drag-and-drop unit processes to build a wastewater treatment plant schematic, automatically calculate a design and then estimate the costs of building, operating and maintaining the facility. At the planning level, current engineering practices primarily use empirical modeling techniques in combination with cost databases. This involves the gathering of historical capital and operating costs of similarly sized plants with similar wastewater and treatment characteristics. These techniques often estimate the cost based on only a single wastewater parameter, such as wastewater flow rate. *CapdetWorks*® is a more comprehensive system, since the design is based on all the characteristics of the wastewater being treated.

CapdetWorks® 2.1 uses both empirical costing models and design algorithms to individual process and pieces of equipment. About 60 treatment processes are provided, including physical/chemical, biological, sludge stabilization, handling and de-watering technologies. From the user's plant layout, the software automatically calculates the required unit process dimensions and equipment. It also allows the engineer to override any of the calculated designs. There is a sophisticated scenario management feature that encourages the user to lay out many treatment alternatives and rapidly calculate and compare costs between them. Capital and operational costs for each process technology can be localized, or users can create their own cost index or apply published industry cost indices to the default values.

CapdetWorks® version 1.0 is available for most Microsoft Windows operating systems; at the time of writing this, it costs US$ 2,450 per license. Free evaluations can be downloaded from the company website at http://www.hydromantis.com.

9.3 Estimating overall costs of wastewater treatment processes with substance and energy recovery

Overall costs of wastewater treatment processes with substance/energy recovery in a treatment facility are the sum of capital costs and operating costs, minus sale price or savings of recovered substances and/or energy. However, forecasting cost savings as a result of recovered substances and/or energy is difficult. Whether a new product or energy from wastewater treatment facility will be accepted in the marketplace depends on several factors, including any additional costs of producing the product, properties of the product, environmental impact, public acceptance, and governmental subsidies.

An additional hurdle to forecasting the fate of a recovered product from food and agricultural wastewater treatment process is that price and/or availability of the competing alternative to the recycled product is also changing constantly. This makes any meaningful long-term forecasting of economical benefits of energy/substance recovery from wastes contentious. Biofuel is a case in point; if the petroleum oil price in the world market goes through the roof, or there is a widespread shortage of petroleum products due to catastrophes or wars in oil-producing nations or regions, then biofuel will be very competitive in price.

9.4 Further reading

Zall, R.R. (2004). *Managing Food Industry Waste: Common Sense Methods for Food Processors*. Blackwell Publishing Professional, Ames, IA, USA.

Qasim, S.R. (1998). *Wastewater Treatment Plants: Planning, Design, and Operation*, Second Edition. CRC Press, Boca Raton, FL, USA.

Vesilind, P.A. (2003). *Wastewater Treatment Plant Design*. Water Environment Federation & IWA Publishing, Alexandria, VA, USA.

9.5 References

Brown, R.C. (2003). *Biorenewable Resources: Engineering New Products from Agriculture*. Blackwell Publishing Professional, Ames, IA, USA.

Peters, M.S. & Timmerhaus, K.D. (1980). *Plant Design and Economics for Chemical Engineers*. 3rd Edition. McGraw-Hill, New York, NY.

Tsagarakis, K. P., Mara, D. D. & Angelakis, A. N. (2004). Application of cost criteria for selection of municipal wastewater treatment systems. *Water, Air, and Soil Pollution* **142**, 187–210.

Ulrich, G.D. (1984). *A Guide to Chemical Engineering Process Design and Economics*. Wiley, New York, NY.

Uluatam, S.S. (1991). Cost models for small wastewater treatment plants. *International Journal of Environmental Studies* **37**, 171–181.

US EPA (1983). *Construction Costs For Municipal Wastewater Treatment Plants: 1973–1982*. EPA Publication 430983004. Washington, DC, USA.

Wright, D.G. & Woods, D.R. (1993). Evaluation of capital cost data. Part 7: Liquid waste disposal with emphasis on physical treatment. *Canadian Journal of Chemical Engineering* **71**, 575–590.

Wright, D.G. & Woods, D.R. (1994). Evaluation of capital cost data. Part 8: Liquid waste disposal with emphasis on biological treatment. *Canadian Journal of Chemical Engineering* **72**, 342–351.

Wright D.G. & Woods, D.R. (1995). Evaluation of capital cost data. Part 9: Liquid waste disposal with emphasis on chemical treatment. *Canadian Journal of Chemical Engineering* **73**, 546–561.

Index

Absorption, 65
Activated energy, 107:
   Arrhenius correlation, 107
   temperature dependence of reaction rates, 107–8
Activated sludge, 6, 12, 31, 33–6, 38, 52, 57, 116, 118, 121, 125, 158, 160–161, 197–203, 249:
   characteristics of, 196
   types of, 196
   design of, 110–114
Adsorption, 2, 12, 58, 59, 63, 65:
   adsorbate, 65
   adsorbent, 65
   breakthrough curves, 68
   capacities, 66
   chemical adsorption, 65
   definition, 65
   factors influencing, 67–8
   Freundlich, 66–7,
   isotherms, 66
   kinetics, batch reactors, 25
   kinetics, flow reactors, 68
   langmuir, 66–7
   linear, 66–7
   physical processes, 65
   rates, 67
   system design, 68–9
   unfavorable, 66
   van der Waals forces, 59
Aeration, 34, 121, 124, 156, 203, 209:
   diffuse air, 159
   mechanical devices, 157

Aerated lagoon, 114, 116
Aerobic conditions:
   in aquatic systems, 189
   in pond systems, 159
   wetland systems, 182
Aerobic ponds, 159
Agricultural crops:
   function of, 166
   in overland flow management of, 166
Agricultural recycling and reuse, 231, 234 (*see also* recoverable products)
Air stripping:
   of ammonia, 135
   of VOCs, 71
Algae, 29, 30, 34, 35, 115, 148, 237
Alkalinity, 4, 141, 165, 209
Alum:
   flocculant 61
   for phosphate removal, 131, 141, 143
   sludge with, 196, 208
Ammonia, 39–41, 134–7
Anaerobic conditions:
   in aquatic systems, 183, 188
   in biological process, 31, 36, 103, 114, 126–131, 138–40
   in pond systems, 160–161
   in wetland systems, 183
Anaerobic digesters for sludge stabilization, 203–4
Application methods:
   in hyacinth systems, 186
   in duckweed systems, 188
   in land systems, 167

Food and Agricultural Wastewater Utilization and Treatment, Second Edition. Sean X. Liu.
© 2014 John Wiley & Sons, Ltd. Published 2014 by John Wiley & Sons, Ltd.

INDEX

Application range of floating plants, 191
Application rates of wastewaters:
    for land treatment, 169
    for overland flow, 176
    for rapid infiltration, 180
    slow rate, 172
    for sludge systems, 165, 218–219
Aquatic systems:
    description of, 185
    design considerations, 189, 190
    organic removal, 186
    nitrogen removal, 186
    phosphorus removal, 186
    temperature effects, 191
ATP:
    structure of, 37
    synthesis of, 38–9

Bacteria, 9, 15, 30–34, 105:
    in anaerobic contact processes,128
    in biological phosphorus removal, 138
    in land treatment systems,176
    in stabilization pond systems,115
    in sludge treatment, 203
    in trickling filters or biofilters, 117
    in wetland systems, 182
Bacterial kinetics, 22, 40
Biodiesel, 228, 237–9
Biogas, 227, 235–9
Biological nutrient removal, 135, 137
Biological processes, 114–31
Biomass, 106, 108, 115, 118, 124, 127, 131, 154, 157, 166, 187
Bioethanol, 238–9
Bioresource utilization, 228
Bulking, 206

Catalyst, 16, 20, 22, 31, 142, 175
Chemical coagulants, 58–61
Chemical oxidation, 6, 16, 70–71
Chemical potential, 81, 146
Chemical precipitation, 59, 141, 143, 180
Chick's law, 149
Chlorination, 8, 71, 137
Chlorine, 8, 137, 147–8

Coagulation, 57:
    coagulant, 58–63, 142, 196, 208
    coagulant aid, 60 (*see also* polyelectrolyte)
    colloid, 60, 65
    destabilization of colloidal dispersion, 60
    jar test, 61
    polyelectrolytes, 60
    processes, 58–9
    sludges, 57–60
Completely stirred tank reactor (CSTR), 26
Conservation of mass (material balance), 19, 26
Conventional processes, 143, 249
Contactor, 104, 116, 119–25
Corrosion, 78, 143

Darcy's law, 64
Denitrification, 33, 41–2, 134–6, 140
Desal process, 99
Design, 36, 139, 145, 155–65, 176–80, 184–91, 198, 199–200, 206, 213, 247–54
Diatomaceous earth, 63
Diffusion:
    diffusivity, 77
    diffusive mass transfer, 63, 77, 146
    in filtration, 63
    membranes, 73
Discrete settling, 52–3, 55
Disease, 188
Disinfectants:
    calcium hypochlorite (ca(ocl)$_2$), 148
    chemical disinfectants, 148
    chlorine, 148
    chlorine dioxide, 148
    hydrogen peroxide, 148–9
    ozone, 148

Electrical double layer, 58–60
Electrical neutrality, 59
Electrode, 80
Electrodialysis:

INDEX

applications, 72, 79–80
boundary layer, 81
Electrostatic forces (interactions), 63, 57
Energy, activation, 22:
   electrochemical, 74
Enthalpy, 85
Equilibrium, 22, 59, 65, 67, 70, 81, 147
E. Coli, 32

Faraday's constant, 80
Faraday's law, 80
Fenton reaction, 210
Ferric and ferrous iron, 61–2
Filter:
   filter cake, 63–5, 208
   filter media, 63, 144
   filter press, 210–214
   process, 62
   sand, 45, 62
   vacuum, 62–3
Filtration:
   application, 45, 57, 62
   Kozeny-Carmen equation, 78
   Darcy's law, 64
   mathematical models, 64
   mechanisms, 62–3
   precoat, 63–4
   sludge dewatering, 209–14
   underdrainage system, 63
   vacuum (see also vacuum filter)
First order reaction:
   application, 109
   rate equation, 21
   reactions, 21
   reversible reactions, 21
Flocculation:
   collision efficiency, 58
   definition, 56
   jar test, 61
   transport of colloidal particles, 58
Flow reactors, 20
Fluidized bed reactors, 216, 220–221
Flux:
   diffusive mass transfer, 75, 81, 85, 867, 89

membrane separation, 74, 75, 76, 82
solvent (water), 74, 76
Freundlich:
   isotherm (equation), 67
   linearization, 67–8

Gas transfer (oxygen):
   applications, 31–3, 37, 41, 203, 205, 220
   biological systems, 114, 116, 119, 139, 149
   rates, 127
Gas transfer systems:
   compressed air, 217
   mechanical, 114, 119
Granular sand filtration, 45

Hagen-Poiseuille relationship, 76
Head loss, 64
Heavy metal, 134, 186–8, 190, 196–7, 204–6, 240–241
HRAP, 157
Hydrogen, peroxide:
   chemical disinfection, 148
   decomposition, 149
Hydrogen, sulfide, 122–3, 127
Hydrophobic interaction, 66
Hydrophobic membrane, 84–5
Hyperfiltration (see also reverse osmosis)

Incineration, 217
Ion Exchange:
   applications, 80, 93, 135, 137
   chemical properties, 93, 95, 97
   design, 97–8
   equilibria, 94
   system, 99 (see also DESAL)
   ion exchange materials, 97
   membranes, 137
   methods of operations, 98–9
   reactions, 98
   regeneration, 98–9
   synthetic materials, 97
   water demineralization, 98

Ion exchange resins:
  Description, 96
  exchange capacity
    regenerations, 94–5
  strong acidic, 95
  strong basic, 95
  weakly acidic, 96
  weakly basic, 96
Ion exchange systems:
  DESAL process, 99
  design efficiency, 98

Kinetics of reactions, 20–24
Kozeny-Carmen equation, 78

Langmuir equation, adsorption:
  isotherm, 66
  linearization, 67
Lévêque's correlation, 83
Lime:
  oxidation, 70
  sludge conditioning, 199, 201–2, 208
  regenerate zeolites, 137
  removal of nitrogen, 137
  removal of phosphate, 141

Material balances:
  conservation of mass, 18
  enzymatic reactions, 23
  reaction rates, 19
  reactors, 25
Membranes:
  concentration polarization, 74–5, 78, 81, 83, 92–3, 145–6
  flux, 76, 81–2, 85, 146
  fouling, 92
  ion selective, 137
  microfiltration, 77
  modeling, 77–8
  modules, 145
  nanofiltration, 78
  permeability, 64, 74–5
  permselective, 81
  pervaporation, 81
  retention, 73

reverse osmosis, 72, 75, 78
separation processes, 74
solute rejection, 74
ultrafiltration, 78, 144, 230
Michaelis-Menten kinetics, 22–3
Microalgae, 237–9
Microorganisms:
  classifications, 30–35
  denitrification, 41
  Monod model, 109
  nitrification, 39
  nutrient requirements, 36
  role of, 35

Nitrate:
  removal by biological processes, 135–6
  removal by electrodialysis, 80
  removal by ion exchange, 93
  removal by physicochemical processes, 136–7
Nitrite:
  conversion of, 40
  PAOs, 140
  removal by biological processes, 136

Odor, 14
Organic matters:
  removal by adsorption (*see* Adsorption)
  removal by aeration (*see* Aeration)
  removal by bioconversion (see Biological processes)
  removal by coagulation (*see* Coagulation)
  removal by oxidation (*see* Chemical oxidation)
  sludge (*see* Sludge)
  BOD or BOD_5 (Biological Oxygen Demand), 14–15
  COD (Chemical Oxygen Demand), 16
Osmosis:
  Reverse (*see* Membrane)
Osmotic pressure difference, 74
Oxidizing agents, 148

Oxygen:
   chemical oxidation (*see* Chemical oxidation)
   oxidation-reduction (*see* Redox reaction)
Ozone:
   decomposition, 148
   disinfection, 147–8

Packed bed reactors, 68–70
Pathogens, 147–8
Permeability:
   coefficient of, 64
   membranes (*see* Membrane)
   solute, 75
   water/solvent, 74
Pervaporation (*see* Membrane):
   applications, 12, 72, 81, 232–3, 235
   boundary layer, 82
   separation factor, 82
Permselective:
   membranes (*see* Membrane)
Phenol:
   chemical oxidation, 70
   from protein degradation, 32
Phosphate:
   removal by biological processes, 137
   removal by coagulation, 142
   removal by ion exchange (*see* Ion exchange)
   removal by physicochemical processes, 141–6
   removal by precipitation, 142–3
Polarization, concentration (*see* Membrane)
Polyelectrolyte:
   coagulation (*see* Coagulation)
   sludge conditioning, 198, 210
Pore size, 74, 77–8, 84
Porosity, 64
Porous solids:
   membranes separation principle, 72
   silica, 63
   composting, 205
Process design, 18–19, 26, 86
Pyrolysis, 241

Radicals, 71
Rapid mixing, 58, 61
Rate constant, 20–22, 88
Rates of reaction, 20, 22
Reaction rates, see rates of reactions
Reactions:
   biochemical, 103, 105
   chemical, 20–22, 70, 142–9, 210
   enzymatic, 23–4, 105
Reactors:
   batch, 25
   CSTR, 26
   design, 25, 86
   in series, 104
   membrane separation, 88–91 (*see also* membrane module)
   sedimentation (*see also* sedimentation tank)
   sludge thickeners, 198–9
   trickling filters, 117
   redox reaction, 38, 70
   residence time, 27, 53–4, 125
Reverse osmosis:
   applications, 73, 75, 78
   module types, 88
   osmotic pressure, 74, 76
   separation factor, 73
   solute rejection, 74
Reynolds number, 76

Sand, filtration (*see* Filtration)
Sedimentation:
   type-I, 53
   type-II, 55
   coagulation, 57
   compression, 57
   sedimentation tank, 53–5
   sludge, 60
   terminal velocity, 54–5
   zone settling, 53
Silver, 16
Selectivity, ion exchange (*see* Ion Exchange)
Selectivity, membranes (*see* Membrane)
Settling (*see* Sedimentation)
Settling velocity, 53, 55, 57

INDEX

Sherwood number correlations, 76
Sludge:
    characteristics, 196
    chemical properties, 197
    dewatering properties, 196
    fuel values, 219–21
    land applications, 166
    specific gravity, 197–8
Sludge treatment:
    aerobic digestion, 203
    anaerobic digestion, 203–4
    centrifugation, 210
    chemical conditioning, 208
    combustion, 219
    conditioning, 208
    dewatering, 209
    disposal
        by land applications, 166, 218–219
        to surface, 218–219
    drying, 216
    filter pressing, 210, 212–213
    flash drying, 216
    flotation, 197
    fluidized bed, 216
    freezing, 214
    gravity thickening, 196, 198–9
    heat treatment, 208–9
    lagooning, 195
    multiple hearth incineration, 216, 220
    polymers, 208
    rotary dryer, 216
    sludge management, 195, 219
    thickening (concentration), 197
    vacuum filtration, 209
Sodium hypochlorite, 148
Softening, 93, 96
Solubility, 87, 135
Solute permeability, 75
Solute rejection, 74
Sorption, 65, 93
Stability of colloids, 58
Steady state:
    determination of rate parameters, 75
    reactors, 20, 22–3, 26, 113–114

Sterilization:
    heat, 149
    irradiation, 149
Stoichiometry, 18
Stoke's law, 57
Sulfite, ion exchange, 95
Surface area, 78, 84, 89, 114, 121, 125, 162, 177–8, 184–5, 191–2, 204
Suspended solids, 1, 3, 9, 12–13, 45, 48, 50, 52, 55, 57

Temperature, parameter of physicochemical treatment, 13
Terminal settling velocity, 57
Thermodynamics:
    reactions, 24
Tortuosity, 78
TSS (Total Suspended Solid), 9

Ultrafiltration:
    applications, 71, 144–5
    concentration polarization, 74–5, 145–6
    definition, 78
    design, 86
    membrane properties, 72
    (Hagen-)Poiseuille relationship, 77
        (*see also* microfiltration)
    retention, 72–3
    ultraviolet irradiation, 149
Underdrain systems, 63

Van der Waals attraction:
    coagulation, 59, 66
Viruses, 30, 133, 204, 240

Water permeability, 74
Water recovery, 228
Whey protein recovery, 229–30

Zeolites:
    ion exchange, membranes, 93
    adsorption, 66
Zero discharge, 228
Zeta potential, 58